ALGEBRA

Essentials Practice

WORKBOOK

with Answers

Linear & Quadratic Equations, Cross Multiplying, and Systems of Equations

Improve Your Math Fluency Series

$$2x^2 - 3x = -1$$

Chris McMullen, Ph.D.

Algebra Essentials Practice Workbook with Answers:
Linear & Quadratic Equations, Cross Multiplying, and Systems of Equations

Improve Your Math Fluency Series

Updated on March 31, 2014

CreateSpace

Professional & Technical / Science / Mathematics / Algebra
Professional & Technical / Education / Specific Skills / Mathematics / Algebra

ISBN: 1453661387

EAN-13: 9781453661383

Contents

Making the Most of this Workbook

- Mathematics is a language. You can't hold a decent conversation in any language if you have a limited vocabulary or if you are not fluent. In order to become successful in mathematics, you need to practice until you have mastered the fundamentals and developed fluency in the subject. This *Algebra Essentials Practice Workbook with Answers: Linear & Quadratic Equations, Cross Multiplying, and Systems of Equations* will help you improve the fluency with which you apply fundamental techniques to solve for unknowns.

- This book is conveniently divided into 7 chapters so that you can focus on one basic skill at a time. The exercises of Chapters 1 and 2 are linear equations with a single unknown, which entail using arithmetic operations to collect and isolate the unknowns. The coefficients are all integers in Chapter 1, while Chapter 2 features fractional coefficients. Quadratic equations are the focus of Chapters 3 thru 5. Chapter 3 is limited to quadratic equations that either do not have a constant term or do not have a term of the first power in the unknown, while Chapters 4 and 5 are focused on techniques for solving the full quadratic equation. The quadratic equations of Chapter 4 are reasonably straightforward to factor into a product of linear factors with integral coefficients, whereas the focus of Chapter 5 is how to apply the quadratic formula. Chapter 6 is dedicated to cross multiplying. Systems of linear equations with two unknowns are the subject of Chapter 7.

- Each chapter begins with a few pages of instructions describing a basic algebraic skill – such as how to factor the quadratic equation. These instructions are followed by a few examples. Use these examples as a guide until you become fluent in the technique.

- After you complete a page, check your answers with the answer key in the back of the book. Practice makes permanent, but not necessarily perfect: If you practice making mistakes, you will learn your mistakes. Check your answers and learn from your mistakes such that you practice solving the problems correctly. This way your practice will make perfect.

- Math can be fun. Make a game of your practice by recording your times and trying to improve on your times, and recording your scores and trying to improve on your scores. Doing this will help you see how much you are improving, and this sign of improvement can give you the confidence to succeed in math, which can help you learn to enjoy this subject more.

Chapter 1: Linear Equations with Integral Coefficients

Let's begin with a few basic definitions. You can do algebra without knowing what these words mean, but this vocabulary is quite useful for discussing how to solve equations. For example, if your teacher tells you to divide both sides of an equation by the coefficient of x, you won't understand what to do unless you know what a coefficient is.

- The **terms** of an equation are separated by + and – signs. For example, there are 4 terms in the equation $5x - 3 = 9 + 2x$. These include the 5 x, the 3, the 9, and the 2 x.
- A **variable** is an unknown that is represented by a symbol, such as x. For example, there are 2 variables in the equation $3x - 2 + 4y = 3x + 6$. These are x and y.
- A **constant** is a number that is not a variable. For example, the 3 is a constant in the equation $y - 3 = x$.
- A constant that multiplies a variable is called a **coefficient**. For example, there are 2 coefficients in the equation $3x - 4 = 2x$. The coefficients are the 3 and the 2.
- An equation is **linear** if it consists only of constant terms and terms that are linearly proportional to the variables; if any variable is raised to a power, like x^4, the equation is not linear. As examples, $4x - 2 = 3$ is a linear equation, while $5x^3 - 7x = 9$ is not.

The exercises of this chapter consist of linear equations with a single variable. The coefficients and other constants of this chapter are all integers. Now we will discuss some fundamental concepts. It is important to grasp these concepts in order to understand what you are doing – in addition to being able to do it. We will also describe how to apply these concepts to solve such linear equations.

The left sides and right sides of an equation are equal. For example, $3 = 1 + 2$ expresses that 3 is the same as 1 plus 2. Similarly, $4x = 8 + 2x$ means that – whatever x is – if you add 8 to 2 times x, it will be equal to 4 times x. Since the two sides of an equation are equal, if you perform the same mathematical operation (like addition or multiplication) to both sides of the equation, it will still be true.

Here is an example of what this means. Consider the simple equation, $3 = 3$. If you add 7 to both sides of the equation, as in $3 + 7 = 3 + 7$, you get $10 = 10$, which holds true. If instead you multiply both sides of the equation by 4, you get $(3)(4) = (3)(4)$, or $12 = 12$. As long as you add (or subtract) the same amount to both sides of the equation – or multiply both sides by the same factor – equality will still hold.

Therefore, you may perform the same mathematical operation to both sides of an equation. For example, in the equation $2x + 3 = 9$, you could subtract 3 from both sides, obtaining $2x = 9 - 3$, to find that $2x = 6$. Similarly, you could divide both sides of $2x = 6$ by 2 to see that $x = 3$. As a check, $2(3) + 3 = 6 + 3 = 9$, so we see that $x = 3$ solves the equation.

There are a couple of simple rules for what to do to both sides of a linear equation in order to solve for the variable. The main principle can be summarized in three words: Isolate the unknown. This means to first move all of the unknowns to one side of the equation and all of the constant terms to the other, to collect all of the unknowns in a single term, and then to divide by the coefficient of the unknown. Below, we describe the precise steps of the strategy:

- First, decide whether to put your unknowns on the left or right side of the equation. You want the constant terms on the opposite side.
- If there are any constant terms on the wrong side, move them over to the other side as follows: Subtract a positive constant from both sides of the equation, but add a negative constant to both sides of the equation. For example, in $x + 2 = 8$, the 2 is positive, so subtract 2 from both sides to obtain $x = 8 - 2$. However, in $x - 4 = 6$, the 4 is negative, so add 4 to both sides to obtain $x = 6 + 4$.
- If there are any variables on the wrong side, move them over with the same technique. For example, in the equation $3\,x = 4 + 2\,x$, the $2\,x$ is positive, so we subtract $2\,x$ from both sides, obtaining $3\,x - 2\,x = 4$. In comparison, in $x = 12 - 3\,x$, the $3\,x$ is negative, so we add $3\,x$ to both sides to get $x + 3\,x = 12$.
- Once all of the unknowns are on one side and all of the constant terms are on the other, collect the terms together. For example, the equation $5\,x + 3\,x - 2\,x = 7 - 4 + 3$ simplifies to $6\,x = 6$ (since $5 + 3 - 2 = 6$ and $7 - 4 + 3 = 6$).
- Now divide both sides of the equation by the coefficient of the unknown. For example, if $4\,x = 24$, divide both sides by 4 to find that $x = 24 / 4 = 6$. If the unknown is negative, multiply both sides of the equation by -1. For example, if $-x = 4$, turn this into $x = -4$. Similarly, if $-x = -8$, then $x = 8$ because $(-1)(-1) = 1$.
- If your answer is a fraction, check to see if it can be reduced. For example, 6/9 can be reduced to 2/3. You reduce a fraction by finding the greatest common factor in the numerator and denominator. For 6/9, the greatest common factor is 3 because $6 = (2)(3)$ and $9 = (3)(3)$. That is, both the 6 and 9 have the 3 in common as a factor. For 12/24, the greatest common factor is 12 since $24 = (12)(2)$. In order to reduce a fraction, divide both the numerator and denominator by the greatest common factor. For 12/24, dividing 12 and 24 both by 12 results in 1/2. Therefore, $12/24 = 1/2$.

We first illustrate this technique with one example with a detailed explanation of the steps. The remaining examples have no explanation so that you may concentrate purely on the math. Take some time to study these to understand the concepts and, if necessary, reread the preceding text or the first example until you understand the solutions. Once you understand the solutions to these examples, you are ready to practice the technique yourself. You may need to refer to the examples frequently as you begin, but should try to solve the exercises all by yourself once you get the hang of it. Be sure to check the answers at the back of the book to ensure that you are solving the problems correctly.

Example 1: $7x - 4 = 5x + 6$. Let's move the 4 to the right and the 5 x to the left so that all of the variables will be on the left and all of the constant terms will be on the right. Since the 4 is negative, we add 4 to both sides: $7x - 4 + 4 = 5x + 6 + 4$, which reduces to $7x = 5x + 10$. Since the 5 x is positive, we subtract 5 x from both sides: $7x - 5x = 5x + 10 - 5x$ or $2x = 10$. Now we divide both sides by the coefficient of x, which is 2: This gives $2x / 2 = 10 / 2$, which simplifies to $x = 5$. It is a good habit to check that the answer solves the original equation by plugging it back in: $7(5) - 4 = 35 - 4 = 31$ and $5(5) + 6 = 25 + 6 = 31$. Since both sides are equal to 31, we have verified that $x = 5$ is indeed a solution to the equation.

Example 2: $9 + 4x = x$.

$$9 + 4x = x$$
$$9 + 4x - 4x = x - 4x$$
$$9 = -3x$$
$$9 / (-3) = -3x / (-3)$$
$$\boxed{-3 = x}$$

Check: $9 + 4(-3) = 9 - 12 = -3$. ✓

Example 3: $5 - 2x = 4x - 7$.

$$5 - 2x = 4x - 7$$
$$5 - 2x + 7 = 4x - 7 + 7$$
$$12 - 2x = 4x$$
$$12 - 2x + 2x = 4x + 2x$$
$$12 = 6x$$
$$12 / 6 = 6x / 6$$
$$\boxed{2 = x}$$

Check: $5 - 2(2) = 5 - 4 = 1$.
$4(2) - 7 = 8 - 7 = 1$. ✓

Example 4: $2x + 5x = 3x$.

$$2x + 5x = 3x$$
$$7x = 3x$$
$$7x - 3x = 3x - 3x$$
$$4x = 0$$
$$4x / 4 = 0 / 4$$
$$\boxed{x = 0}$$

Check: $2(0) + 5(0) = 0$.
$3(0) = 0$. ✓

Example 5: $7 = 2 - x$.

$$7 = 2 - x$$
$$7 - 2 = 2 - x - 2$$
$$5 = -x$$
$$5 / (-1) = -x / (-1)$$
$$\boxed{-5 = x}$$

Check: $2 - (-5) = 7$. ✓

Here are a few more notes:

- Multiplication is often expressed with a cross (×), but this symbol is easily confused with x in algebra. Therefore, in algebra, multiplication between two numbers is usually represented with a dot (·) or consecutive parentheses. For example, $5 \cdot 9 = 45$, $(7)(3) = 21$, and $5(4) = 20$.
- Adding a negative number equates to subtracting a positive number, and subtracting a negative number equates to adding a positive number. For example, $4 + (-3) = 4 - 3 = 1$ and $7 - (-2) = 7 + 2 = 9$.
- When multiplying two numbers, if one factor is negative, the product is negative, but if both factors are negative, the product is positive. For example, $(-6)(4) = -24$, $(2)(-8) = -16$, and $(-3)(-5) = 15$. Similarly, $16 / (-4) = -4$, $-28 / 7 = -4$, and $-18 / (-9) = 2$.
- Note that equations are reversible – i.e. if $3x = 9$, it is also true that $9 = 3x$.

❶ $9 - 2x = -8 - 4x$

❷ $9 + 8x = 5x + 6$

❸ $2x + 5x = -2$

❹ $-3x - 8 = -x - 4$

❺ $9x - 1 = -5x$

❻ $-3x - 5x = -8 - 4x$

❼ $-3 + 4x = -5 - 6x$

❽ $2x + 6 = 5x - x$

❾ $-4x + 2x = 4 + 4x$

❿ $-9 + 5x = 7 - 5x$

⓫ $3 + 2x = 5 + 4x$

⓬ $7x + 7x = -3 + x$

⓭ $-3x + 9x = -5 + 5x$

⓮ $-1 = 5 + 7x$

⓯ $4x + 8x = 9 - 6x$

⓰ $-8x - 6x = -2x - 7$

⓱ $7x + x = 7x - 4$

⓲ $9x + 3x = 7x + 5$

❶ $-9x + 4 = 7 - 8x$

-3/11

❷ $-3 + 7x = 6 - 9x$

-.45=x

❸ $-2 = -x - 5x$

1/3

❹ $-4 - 3x = -5x + 6$

5=x

❺ $x - 5 = 8x + 9x$

5/17

❻ $9 - x = 4x - 6x$

x=-9

❼ $7x + 7 = 2x - 8$

-3=x

❽ $-8x + 5x = 3 + 2x$

-.6=x

❾ $7x + 8 = 5x + 4$

x=-2

❿ $-9x - 5 = -3x + 2$

-4/9

⓫ $-5x = 4 + 8x$

4/15

⓬ $-3x = -8x$

x=0

⓭ $6x + 4x = 3x + 8$

-6/7

⓮ $-3x = -8 + 2x$

1.6=x

⓯ $-7 - 6x = 3x + 8$

16/9

⓰ $-6x - 4 = -2x - 9$

1.25=x

⓱ $4 - 9x = 9x - 2$

1/3

⓲ $8x = -4x - 3x$

x=0

1 $5 + 9x = 5x - 4$

1 5/19

2 $3 - 5x = 3x - 2x$

x=5

3 $-6x + 2x = -1 - 9x$

-.2=x

4 $x + 7 = -6 + 3x$

6.5=x

5 $7x - 4 = x$

4/6

6 $4x - 8 = -7x - 2$

4/11

7 $3x + 8 = 3x + 8x$

8/7

8 $-2x - 4 = -9 + 5x$

5/7

9 $-2x - 6 = 2x + 6x$

-.6=x

10 $-8x - 3x = -6 + 5x$

x=0

11 $9x + 6x = -3 + 7x$

x=.375

12 $8x - 1 = 9 + 3x$

2=x

13 $-x - 8 = -9x - 2$

x=.75

14 $-5x + 2x = -8 - 5x$

1=x

15 $4 = 9x + 5$

1/9

16 $5x + 5x = 5x + 4$

.8=x

17 $1 - 4x = 8x - 4x$

0=x

18 $9x - 8 = -4x + 3x$

.8=x

1. $4 + x = -9x + 1$

   $-.3 = x$

2. $4 + 6x = 5x - 8$

   $-12 = x$

3. $-6x + 9x = -4x + 1$

   $-1 = x$

4. $8x - 2 = 5 - x$

   $-7/-9$

5. $-8 - 2x = 2x - x$

   $2\,2/3$

6. $-7x + 2x = -4$

   $-1.25 = x$

7. $-3 - 5x = x + 3$

   $-1 = x$

8. $2x - 4x = 9 + 8x$

   $x = -.9$

9. $3 - 8x = 2x + 7x$

   $3/19$

10. $1 - 5x = 5 + 8x$

    $4/13$

11. $9 - 3x = 8x - 3x$

    $4.5 = x$

12. $-7x = 4 + x$

    $-.5 = x$

13. $-6x + 4x = -8x - 6$

    $x = -1$

14. $-3 + 5x = -8x$

    $3/13$

15. $3x - 9x = -9 + 8x$

    $9/14$

16. $-8x + 7 = -5x - 6$

    $4\,1/3$

17. $-8x + 9 = 2x + x$

    $.9 = x$

18. $9x - 3x = 7x + 4$

    $-.25 = x$

❶ $9x - 8x = 4x + 8$

−.375=x

❷ $-x + 2 = -1 + 2x$

1=x

❸ $-4 = -6x - 1$

.5=x

❹ $2 + 2x = 4x + 8$

3/6/2

❺ $3 + 5x = 6x + 1$

.5=x

❻ $-x - 2x = 3x$

0=x

❼ $x + 7x = -3x + 6$

6/11

❽ $-x + 6x = 9x - 7$

−1.75=x

❾ $2x - 5 = -3 + 6x$

.5=x

❿ $-5 - 6x = 8$

−1.25=x

⓫ $-3x - 2x = -7x + 1$

1/12

⓬ $3x - 5 = -2 + 7x$

−.75

⓭ $-4x - 3 = -5x + 8$

1/11

⓮ $6x + x = -5 + 2x$

−5/9

⓯ $2x - 7x = -2x + 6$

−1.5=x

⓰ $-7x - 9 = -x - 7$

2/6

⓱ $-4 + x = -3 - 4x$

.2=x

⓲ $5 - 6x = 7 - 3x$

2/9

❶ $-x + 3x = -8x - 1$

.1=x

❷ $-8x + 7 = -3x + 8$

.2=x

❸ $2x + 7 = 9x + 3x$

.12=x

❹ $-2x + 6x = 7x + 5$

-5/3

❺ $-9 = -1 + 4x$

2=x

❻ $-9x - 6x = x - 2$

.125=x

❼ $3 + 2x = 4x - 8$

5.5=x

❽ $-x + 2 = -6x + 4$

.4=x

❾ $-x - 5x = -2x$

0=x

❿ $-1 - 4x = 6 + x$

1=x

⓫ $2 - x = 5x - 3x$

2/3

⓬ $5 = -3x + 2x$

-5=x

⓭ $-7x - 9x = 3x$

0=x

⓮ $7x - 6 = x + 5$

11/6

⓯ $-x - 2 = 4 - 9x$

.75=x

⓰ $3x + 3x = -4 - 7x$

-4/13

⓱ $x = -3x - 6x$

0=x

⓲ $-5 = 3x$

5/3

❶ $9 - 6x = 3x - x$

8/9

❷ $-5x + 8 = -5 - 6x$

3=x

❸ $-8 = 8x + 5x$

8/13

❹ $9 + 7x = 3x + 9$

0=x

❺ $4x + 6x = 9x - 2$

-2=x

❻ $-4x - 8x = 6x + 1$

-1/18

❼ $-6x - 2 = -6x - 4x$

.5=x

❽ $8 + 7x = -8 + 9x$

1=x

❾ $-5x - 9x = -3x - 6$

6/11

❿ $4x = -2x + 5$

1.2=x

⓫ $-3x + 3x = 7x + 2$

-2/7

⓬ $2x - 5 = -7 - 9x$

-2/11

⓭ $4 + 5x = 2x - 9$

.6=x

⓮ $-4 - 2x = 4x$

4/6

⓯ $-4 - 9x = -9x - 3x$

-4/3

⓰ $4x + 5x = 3 + 7x$

1.5=x

⓱ $-9x - 7 = 3x - 6x$

-7/6

⓲ $2 - 8x = 1 - 5x$

-1/3

❶ $-9\,x - 8\,x = 1 + 9\,x$

1/26

❷ $9 + 7\,x = 4\,x + 2\,x$

-1/9

❸ $2\,x + 2\,x = -4 - 8\,x$

3=x

❹ $-6\,x = x - 6$

-7/6

❺ $-2\,x + 6\,x = 3\,x$

0=x

❻ $8\,x + 3 = -6\,x + 8$

2.8=x

❼ $4\,x = -4\,x + 2\,x$

0=x

❽ $5 = -7\,x$

5/7

❾ $3 + 3\,x = -7\,x + 7\,x$

-1=x

❿ $-9 = -4 - x$

-5=x

⓫ $-1 + 9\,x = -5\,x$

1/14

⓬ $-4 + 3\,x = 2\,x - 4\,x$

-1.25

⓭ $-3\,x - 6\,x = 4\,x - 7$

-13/7

⓮ $-7 + 4\,x = -4 + 8\,x$

0=x

⓯ $4\,x - 9\,x = -2$

-2.5

⓰ $-7 + x = -4\,x + 1$

.5=x

⓱ $4 + 3\,x = 9\,x + 9$

-1/3

⓲ $-5 + 4\,x = -x - 4$

.2

❶ $3 + 4x = -5x - 8$

-5=x

❷ $3x + 1 = -6 - x$

1.75=x

❸ $8x = 6 - x$

6/9

❹ $-3x + 5x = 3 + 5x$

-1=x

❺ $4x + x = -8x + 9$

9/13

❻ $-7x - 9 = -1 + 3x$

.8=x

❼ $x + 1 = 3x - 9x$

-1/7

❽ $4 = 9 - 8x$

-.625=x

❾ $-3x = -7x - 6x$

0=x

❿ $-5 + 6x = -8x + 4$

-9/14

⓫ $4x + 5 = 2$

-.75=x

⓬ $2x - 3x = -4x + 3$

1=x

⓭ $2x = -7x + 5x$

0=x

⓮ $-9 + 4x = x + 2x$

9=x

⓯ $6x + 7 = -6$

6/-13

⓰ $-1 - 2x = -4x + 2$

1.5=x

⓱ $-4 - 2x = -5x$

-4/3

⓲ $8 = -8x - 3x$

8/-11

❶ $2 + 4x = -x - 8$

$-.6 = x$

❷ $-x + 3 = -5x - 7$

$.4 = x$

❸ $x - 2x = 9 + x$

$-2/9 = x$

❹ $7x = -7$

$1 = x$

❺ $7x - 5x = -2 - 8x$

$-.2 = 10$

❻ $8 - 7x = -1 + 2x$

$1 = x$

❼ $8 - 2x = 7x + 1$

$9/7 = x$

❽ $-x + 8 = -6x$

$.625 = x$

❾ $3x - 6x = -1$

$3 = x$

❿ $6 - 3x = 6x - 2x$

$4/6 = x$

⓫ $4 + 9x = -8 + 4x$

$-3.25 = x$

⓬ $-9x + 4 = -9 + 4x$

$-1 = x$

⓭ $4 = 1 + 3x$

$.6 = x$

⓮ $-7x + 4x = -4 - 6x$

$2.25 = x$

⓯ $6 + 6x = -2x + 2x$

$-1 = x$

⓰ $4x = 2x - 4x$

$0 = x$

⓱ $3x + 9x = -2 - 8x$

$.1 = x$

⓲ $9x + x = 1 + 4x$

$6 = x$

1. $8x - 5x = 9x - 7$
2. $2 = -6 - 8x$
3. $-5 - 7x = -9x + 5$
4. $2x + x = -3x$
5. $2x - 8x = 9 + 5x$
6. $7x - 9x = -9 + 2x$
7. $2x + 2 = 3x$
8. $-3x + 4 = 9 - 4x$
9. $-4 - 2x = 6x - 4x$
10. $-6x + 5 = 4x - 8x$
11. $-4 + 3x = 2 + 4x$
12. $-6 = -5x$
13. $-9x + 5 = 2x + 1$
14. $8 + 9x = 8 + 4x$
15. $-7x + 6 = 4 + 7x$
16. $-6x - 7x = 3 - x$
17. $7x + 5 = -9x - x$
18. $7x + 9 = -x - 8$

❶ $6x + 3 = -6 - 9x$

$-5 = x$

❷ $4 - 8x = 8x$

$4 = x$

❸ $1 + 6x = 2x - 1$

$2 = x$

❹ $-8x + 4x = 9 - 2x$

$-2/-9$

❺ $4 - 6x = 4x + 4x$

$3.5 = x$

❻ $-5x + 1 = 3x - 7$

$-1 = x$

❼ $-5x + 1 = 4x + 2x$

$-11 = x$

❽ $4x - 6 = 3$

$4/9$

❾ $-8 = -9 + 8x$

$8 = x$

❿ $8x - 4x = -x - 8$

-1.625

⓫ $6x + 4x = 7 - 8x$

$16/7$

⓬ $5x = 5 - 9x$

$2.8 = x$

⓭ $6x + 1 = 8 + 3x$

$3/9$

⓮ $3 + 4x = 2 + 8x$

4

⓯ $4x + 9 = 2x - x$

$3/9$

⓰ $-2x - 6 = -8x + 7$

$6/13$

⓱ $-6x + 1 = -2 + 3x$

$-3 = x$

⓲ $-9 + 7x = 9x - 6$

$2/3$

❶ $6x + 9 = x + 6$

❷ $3 - 2x = x + x$

❸ $8 - 5x = 3x$

❹ $9 + 8x = -5 - 7x$

❺ $6 - 4x = -7x + 6x$

❻ $-4 + 7x = 9x - 4$

❼ $-4 + 3x = -2 + 8x$

❽ $2x = 6 + 8x$

❾ $-x = 3$

❿ $6 - 4x = 1 + 9x$

⓫ $-2 - 8x = 8x + 8$

⓬ $-5x = -8x - 3x$

⓭ $7x + 8 = 4x - 5$

⓮ $-3x - 9x = -9x - 8$

⓯ $9 + 9x = -7 - x$

⓰ $-2 = 4 + x$

⓱ $-7 + 8x = -5x + 2x$

⓲ $-5 - 9x = 3 + 2x$

❶ $-3 + 9x = 8 - 6x$

❷ $3x + 1 = -5$

❸ $-2x = 2 - 9x$

❹ $6 + x = -x + 9x$

❺ $-6x - 5x = 3x - 4$

❻ $3x + 6 = x$

❼ $-8x = 8x + 5$

❽ $9x - 6x = x + 7$

❾ $-5x + 2x = -8 + 2x$

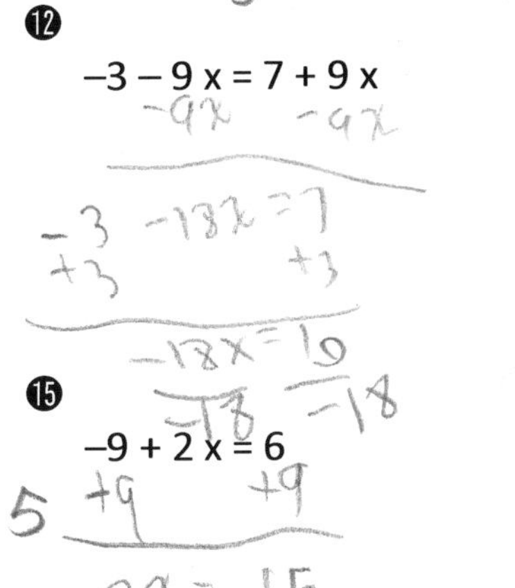

❿ $7 + 4x = -6 - 9x$

⓫ $-7 = -8x + 5$

⓬ $-3 - 9x = 7 + 9x$

⓭ $-7x + 3x = -4$

⓮ $4 = -2x - 5$

⓯ $-9 + 2x = 6$

⓰ $9x - 5x = -7 - 3x$

⓱ $6 + 3x = -8 + 4x$

⓲ $-7x - 8x = 3 + x$

❶ $-8x - 7 = -4x - 7$

❷ $-2x - 5x = -3 + 9x$

❸ $6 - x = -7x - 6$

❹ $1 + 4x = -8$

❺ $x + 3 = 3x - 8x$

❻ $2 - 9x = -8x - 4$

❼ $-x + 9x = 5 + 4x$

❽ $3x + 9 = 2x - x$

❾ $8x + 8 = x - 9x$

❿ $-2 + 3x = -9x - 4x$

⓫ $-5x - 1 = -3x - 5$

⓬ $2x - 8 = 6x + 1$

⓭ $4x + 4x = -4x - 9$

⓮ $-9x + 4 = -5x + 6x$

⓯ $-8x + 3 = 5x - 4x$

⓰ $-1 + 6x = -2x$

⓱ $4 + 5x = 7x - 4x$

⓲ $-1 + 7x = 3x + 5x$

❶ $8x - 3 = -7x - x$

❷ $-5x = -x + 8x$

❸ $-7x = 7x + 5x$

❹ $-9 + 6x = -7x - 6$

❺ $4 - 7x = -2x - 3$

❻ $2x - 1 = 7 - 3x$

❼ $6x + 9x = 4 - 7x$

❽ $-3x - 5x = 6 - 7x$

❾ $-6x - 8x = -x + 4$

❿ $2x + 3x = 3 + 9x$

⓫ $-7x - 6 = -9x - 4$

⓬ $-7x + 2x = 6 - x$

⓭ $-2x + 6x = -7x - 4$

⓮ $7 - x = 5$

⓯ $9 = 2 + 9x$

⓰ $-4x - 8 = 9 + 7x$

⓱ $-7x - 9x = -4x$

⓲ $6x - 9x = 7 + 2x$

❶ $x - 7x = -6 + 9x$

❷ $9x - 1 = 8 - 2x$

❸ $4 + 3x = 2x + 7$

❹ $3 + x = -4x + 7x$

❺ $-1 + 3x = 5x$

❻ $3x - 8 = -7x + 3$

❼ $-3x = 3 + x$

❽ $3x + 3 = -7$

❾ $8 - 4x = -5x - x$

❿ $4 + 7x = -6x - 4$

⓫ $5x - 2 = 2 + 8x$

⓬ $-9x - 2 = -5x - 1$

⓭ $x - 8 = -6x - 2$

⓮ $2 + 2x = 6x + 7$

⓯ $5x - 4 = -7x - 8$

⓰ $-6x + 2x = -2$

⓱ $4x + 8x = 6 - 2x$

⓲ $4x = -8x - 3x$

❶ $-2 + 5x = 8x$

❷ $9x + 8 = -9x + 4x$

❸ $8x + 8 = -2 + 4x$

❹ $9x + 4 = -1 - 2x$

❺ $-8x = -x - 3$

❻ $9x + 9 = -2 - 3x$

❼ $-2 - 5x = 7x + 2$

❽ $-3 = -8x - 4$

❾ $-7x + 6 = -x - 5$

❿ $-4 + 2x = -2x + 5x$

⓫ $x - 9x = 1 - 5x$

⓬ $-2 + 5x = 2x$

⓭ $4 + 6x = -5x - 6x$

⓮ $-2 - 8x = -x - 5$

⓯ $-4x + 4x = -x + 5$

⓰ $-9 - 8x = 2x - 8x$

⓱ $-8x - 5 = 3x + 7x$

⓲ $8x + 1 = 6x - 4$

❶ $1 - 2x = 3 - 5x$

❷ $5 - 2x = -3x - 2$

❸ $-3 + 2x = -7 - x$

❹ $-7 + x = 7$

❺ $6x - 8x = -4x + 6$

❻ $2x - 7 = 4x$

❼ $x - 7x = -4x + 2$

❽ $-4x + 2 = -7 - 3x$

❾ $-9 - 7x = -x + 2x$

❿ $-6 = -5x - 1$

⓫ $-2x - 3 = 3x - 7x$

⓬ $-6x - 8 = 8$

⓭ $8x = 7 - 5x$

⓮ $5x + x = -4 - 2x$

⓯ $7x = 5x - 2x$

⓰ $4x + 4x = -x - 8$

⓱ $-2 + 2x = -9x + 4$

⓲ $-9 - 7x = -4x + 7$

❶ $-3x + 3x = -9x - 8$

❷ $-4x - 3 = -9x - 4$

❸ $5x + 8 = 8 - 6x$

❹ $-4x + 7x = -8x - 9$

❺ $7x + 1 = 6x + 8$

❻ $-7 + 9x = -8x + 8x$

❼ $-9 = 9x + 6x$

❽ $-9x - 8 = -6 - 2x$

❾ $-8 + 3x = 3$

❿ $-8x = -1 + 3x$

⓫ $8x + 3x = -x + 5$

⓬ $-3x - 5 = -x - 9x$

⓭ $2 - 4x = -6x - 2$

⓮ $3x + 2x = 9x + 5$

⓯ $4 + 2x = 5 - 9x$

⓰ $5 - x = -9 + 6x$

⓱ $3x + 5x = -1 + 9x$

⓲ $-6x - 4x = 6 + 5x$

Chapter 2: Linear Equations with Fractional Coefficients

Like the first chapter, the exercises of this chapter consist of linear equations with a single variable. However, many of the coefficients and other constants of this chapter are fractions. These equations are solved using essentially the same strategy, but the arithmetic involves adding, subtracting, multiplying, and dividing fractions.

Let's begin by reviewing the basic concepts associated with the arithmetic of fractions:

- In order to add or subtract two fractions, first find a common denominator. You can find a common denominator by multiplying the two denominators together. For example, (6)(9) = 54 would be a common denominator for 5/6 and 4/9. It is usually convenient to find the lowest common denominator. For 5/6 and 4/9, the lowest common denominator is 18. The lowest common denominator can be found by looking at the factors. The factors of 6 are 2 and 3 and the factors of 9 are 3 and 3. That is, 6 = (2)(3) and 9 = (3)(3). They have the factor 3 in common. The lowest common denominator, (2)(3)(3) = 18, is found by putting the necessary factors from both 6 and 9 together. That is, we need one 2 and two 3's in order to construct both 6 and 9 through multiplication.
- Once you have a common denominator, multiply the numerator and denominator of each fraction by the factor that makes the common denominator. For 5/6 and 4/9, the lowest common denominator is 18. We must multiply 6 by 3 to make 18, so we multiply 5 by 3 to make 15: 5/6 = 15/18. Similarly, we multiply both 4 and 9 by 2: 4/9 = 8/18.
- When both fractions have been changed to have a common denominator you can add or subtract the numerators in order to add or subtract the fractions. The sum 5/6 + 4/9 is found to be 15/18 + 8/18 = 23/18. The difference equals 5/6 – 4/9 = 7/18.
- Sometimes, the resulting fraction can be reduced. For example, if you add 1/2 to 1/3, you get 6/12 + 4/12 = 10/12, and 10/12 can be reduced to 5/6. You reduce a fraction by finding the greatest common factor in the numerator and denominator. For 10/12, the greatest common factor is 2 because 10 = (2)(5) and 12 = (2)(6). As another example, the greatest common factor of 18/24 is 6 since 18 = (6)(3) and 24 = (6)(4). In order to reduce a fraction, divide both the numerator and denominator by the greatest common factor. For 18/24, dividing 18 and 24 both by 6 results in 3/4. Therefore, 18/24 = 3/4.
- In order to multiply two fractions, simply multiply the numerators and denominators together (so it's actually easier to multiply fractions than it is to add them). For example, (2/3)(1/4) = 2/12, which reduces to 1/6. Similarly, (4/9)(3/2) = 12/18 = 2/3.
- In order to divide two fractions, reciprocate (for example, 25/7 is reciprocal to 7/25) the divisor and then multiply the dividend with the reciprocated divisor. (In 36 ÷ 4 = 9, 36 is the dividend, 4 is the divisor, and 9 is the quotient.) For example, 3/4 ÷ 1/6 = (3/4)(6/1) = 18/4 = 9/2. Similarly, 3/2 ÷ 5/8 = (3/2)(8/5) = 24/10 = 12/5.

Here are a few examples illustrating the arithmetic of fractions:

Example 1: 3/8 + 5/12 = ?

$$\frac{3\cdot 3}{8\cdot 3}+\frac{5\cdot 2}{12\cdot 2}=\frac{9}{24}+\frac{10}{24}=\frac{19}{24}$$

Example 2: 3/4 – 4/7 = ?

$$\frac{3\cdot 7}{4\cdot 7}-\frac{4\cdot 4}{7\cdot 4}=\frac{21}{28}-\frac{16}{28}=\frac{5}{28}$$

Example 3: 3/8 × 2/9 = ?

$$\frac{3\cdot 2}{8\cdot 9}=\frac{6}{72}=\frac{6/6}{72/6}=\frac{1}{12}$$

Example 4: 3/4 ÷ 9/8 = ?

$$\frac{3}{4}\div\frac{9}{8}=\frac{3}{4}\times\frac{8}{9}=\frac{3\cdot 8}{4\cdot 9}=\frac{24}{36}=\frac{24/12}{36/12}=\frac{2}{3}$$

The strategy for solving the algebraic equations of this chapter is the same as the strategy from Chapter 1. The only difference is that the arithmetic involves fractions.

- Choose which side of the equation will correspond to variables and constant terms.
- Move constant terms over to the desired side: Subtract a positive constant from both sides, but add a negative constant to both sides. For example, 3x – 1/2 = 3/4 becomes 3x = 3/4 + 1/2 = 3/4 + 2/4 = 5/4.
- Move variable terms over to the other side using the same technique. For example, x/3 = 1/6 – x/4 becomes x/3 + x/4 = 4x/12 + 3x/12 = 7x/12 = 1/6.
- Collect terms together. For example, 2x/3 – 2x/5 = 2/9 + 1/7 becomes 10x/15 – 6x/15 = 14/63 + 9/63, which simplifies to 4x/15 = 23/63.
- Divide both sides of the equation by the coefficient of the unknown. If the coefficient is a fraction, this equates to multiplying both sides of the equation by its reciprocal. For example, 4x/3 = 24 becomes x = 24(3/4) = 72/4 = 18.

Example 1: 5x/2 – 3/4 = 5/8 + x. All of the constant terms will be on the right side if we move the 3/4 to the right. Since 3/4 is negative, we add 3/4 to both sides of the equation: 5x/2 = 5/8 + x + 3/4. We add 5/8 to 3/4 using a common denominator of 8: 5/8 + 3/4 = 5/8 + 6/8 = 11/8. Thus, 5x/2 = 11/8 + x. Now we subtract x from both sides to get all of the variable terms on the left: 5x/2 – x = 11/8. In order to subtract x from 5x/2, we need a common denominator of 2: 5x/2 – 2x/2 = 3x/2. The equation becomes 3x/2 = 11/8. Multiply both sides of the equation by 2: 3x = (11/8)2 = 22/8 = 11/4. Now divide both sides by 3: x = 11/12.

Example 2: $4/3 - x/6 = 3x/4$.

$$4/3 = 3x/4 + x/6$$
$$4/3 = 9x/12 + 2x/12$$
$$4/3 = 11x/12$$
$$(4/3)12 = 11x$$
$$16 = 11x$$
$$\boxed{16/11 = x}$$

Check: $4/3 - (16/11)/(6) = 4/3 - 16/66$
$= 88/66 - 16/66 = 72/66 = 12/11$.
$3(16/11)/4 = 12/11$. ✓

Example 3: $7x/2 = 4x/3 - 3x/4$.

$$0 = 4x/3 - 3x/4 - 7x/2$$
$$0 = 16x/12 - 9x/12 - 7x/2$$
$$0 = 7x/12 - 7x/2$$
$$0 = 7x/12 - 42x/12$$
$$0 = -35x/12$$
$$0 = -35x$$
$$\boxed{0 = x}$$

Check: $7(0)/2 = 0$.
$4(0)/3 - 3(0)/4 = 0$. ✓

Example 4: $4 - x/2 = x/3 - 2$.

$$4 - x/2 + 2 = x/3$$
$$6 - x/2 = x/3$$
$$6 = x/3 + x/2$$
$$6 = 2x/6 + 3x/6$$
$$6 = 5x/6$$
$$(6)(6) = 5x$$
$$36 = 5x$$
$$\boxed{36/5 = x}$$

Check: $4 - (36/5)/2 = 20/5 - 18/5 = 2/5$.
$(36/5)/3 - 2 = 36/15 - 30/15 = 6/15$
$= 2/5$. ✓

Example 5: $2/5 + x/3 = 2/3$.

$$x/3 = 2/3 - 2/5$$
$$x/3 = 10/15 - 6/15$$
$$x = (4/15)\,3$$
$$x = 12/15$$
$$\boxed{x = 4/5}$$

Check: $2/5 + (4/5)/3 = 2/5 + 4/15$
$= 6/15 + 4/15 = 10/15 = 2/3$. ✓

Here are a few more notes:

- Note that something like $2(3/4)/6$ reads as 2 times 3/4 divided by 6, and can be expressed as $(2/6)(3/4) = 6/24 = 1/4$. Similarly, compare $(2/3)6$, which reads as 2/3 times 6, which equals $(2 \cdot 6)/3 = 12/3 = 4$ to $(2/3)/6$, which reads as 2/3 divided by 6, which equals $2/(3 \cdot 6) = 2/18 = 1/9$.
- In the process of isolating the unknown, we repeatedly do the opposite. For example, in the equation $2 + 3x/2 - 4 = 6$, when we add 4 to both sides we are doing the opposite of subtracting 4. Then in $2 + 3x/2 = 10$, when we subtract 2 from both sides we are doing the opposite of adding 2 to $3x/2$: $3x/2 = 8$. When we multiply both sides by 2, we are doing the opposite of dividing both sides by 2: $3x = 16$. Finally, we do the opposite of multiplying both sides by 3, which means to divide by 3: $x = 16/3$.
- If you want to combine fractions that have variables, use the same rules for finding the common denominator as you would for ordinary fractions. For example, to add $3x/4$ to $5x/2$, multiply $3x$ and 4 by 2 and $5x$ and 2 by 4 to make a common denominator of 8: $3x/4 + 5x/2 = 6x/8 + 20x/8 = 26x/8 = 13x/4$.
- Another way to look at $3x/4 + 5x/2$ is by factoring. That is, x is a common factor to $3x/4$ and $5x/2$. If you factor an x out of $3x/4 + 5x/2$, you get $(3/4 + 5/2)x$. Since $3/4 + 5/2 = 13/4$, $(3/4 + 5/2)x = 13x/4$. Here are a few more examples of factoring algebraic expressions: $3x + 8x = (3 + 8)x$, $6x + 4 = 2(3x + 2)$, $x/2 + 1/4 = (2x + 1)/4$, and $-x - 3 = -(x + 3)$. We will discuss factoring in more detail in Chapters 3 and 4. You can complete the exercises of Chapter 2 without any knowledge of factoring.

❶ $-1/2 - 4x/5 = 1/2 - 2x$

❷ $2x - 3x/2 = -x + 1/3$

❸ $5/3 - x/5 = -1$

❹ $-2x - 4 = -1/5 + 3x/4$

❺ $-6x + 1 = x - 1$

❻ $-3x/2 + 1/3 = -1/5$

❼ $x/5 + x/2 = -1/5 - 3x/2$

❽ $-4x/3 - 1/2 = 4x/3 - 2$

❾ $-x - 2x = -x/3 - 3/2$

❿ $-6x - 3 = 5x/4 - 3/4$

⓫ $-5x + 2x/3 = 3/2 - x$

⓬ $5/6 - x = 4x + 6$

⓭ $4/3 - x/3 = -5x/6 + 3x/4$

⓮ $3 + 4x/3 = -2 - 5x/2$

⓯ $-4 + x/2 = -3x + 1/6$

⓰ $-1/2 + 2x = 5/6$

⓱ $2/3 = x + 1/2$

⓲ $5/4 = 3x + x$

❶ $-x + 4/3 = x - 1/2$

❷ $6x + x = -x + 1$

❸ $2/5 = -5/2 - 2x/3$

❹ $1 = 1/5 + 2x$

❺ $-x/5 - 3x/2 = -x/5 + 5/3$

❻ $6/5 + x = -2x + x$

❼ $3x/2 = -5/4$

❽ $2 = 1 + x/2$

❾ $-x/2 - 1/4 = 3x - 3$

❿ $-2x + 3/2 = -1/2 - 2x/3$

⓫ $-2x = 2x - 3x$

⓬ $4/3 - x/2 = -x/6 - 6x/5$

⓭ $-x/2 - 5/3 = 1/3 + x$

⓮ $-1 + x = x/2 + 1/4$

⓯ $-1 + 3x/4 = -x/2 - 1$

⓰ $-x/5 + 6 = -1/3 + 2x/3$

⓱ $-1/3 = 3x + 2x/3$

⓲ $-x + 2x = 5/4 - x/3$

$\frac{15}{16}$

$+\frac{x}{3}$

$-x + \frac{x}{3} + 2x = \frac{5}{4}$

$x + \frac{x}{3}$

❶ $-2 - 2x/3 = 1 - 2x$

❷ $-x = 3x$

❸ $4 + 2x = -x/2 + 3x/2$

❹ $x + x = -4 - 4x/3$

❺ $1 = 3x/2 - 1$

❻ $-x/2 = -x/3 - 2x/3$

❼ $-1/5 + x/4 = -x + 3x/2$

❽ $5x/6 = -3/2 + x/5$

❾ $-2x/5 = 3x - 1/4$

❿ $5x = -3x - 2/5$

⓫ $2/3 = 1 + 2x/3$

⓬ $-2x/3 + 2x/3 = -6/5 - x$

⓭ $-1/6 - x/4 = -5x/4 + 2/3$

⓮ $-3/4 + 5x/4 = -2x/5 - 1/3$

⓯ $6x/5 = -x/2 - 1$

⓰ $x - 3x = -2 + x$

⓱ $2/5 + x/4 = 5/6$

⓲ $-2x/3 + 5 = 3x/2 + 2x$

❶ $-x/2 + 6 = -1/2 + 2x$

❷ $-2x/5 = -1 - x/2$

❸ $1 - 2x = -3x + x/3$

❹ $3/2 - 6x = -6x - 6x/5$

❺ $6x - 1 = 3/5$

❻ $2/5 = -5x - 6$

❼ $4x/3 - 3 = -3 - 5x/4$

❽ $4x/3 = 2x + 1$

❾ $5x/4 - 3/2 = 3x - 5/6$

❿ $-4x - 3x/4 = 3/2 - x$

⓫ $3x = 4/5 + x/4$

⓬ $3 = 1/4 - 2x$

⓭ $-x - 4x = 2x$

⓮ $-x + 2/5 = -3/2$

⓯ $-2x/5 - 6 = -x + 2$

⓰ $x/2 + 1/3 = x/3 + 3/5$

⓱ $-x + 5/4 = -x/4 + 1$

⓲ $5/4 = 6x/5$

❶ $-x/5 + 2x/3 = 3x/5$

❷ $x/3 - 2/3 = 4x - x/4$

❸ $2/3 = 4/5 - 2x/5$

❹ $-x + 3 = x/2 - 2/3$

❺ $-3x/5 - 1/4 = -4x + 6x/5$

❻ $6x/5 - 1/2 = 2/5$

❼ $x/2 + 2x/3 = 6x/5$

❽ $1 - 2x/3 = 3x/2$

❾ $5/6 - 2x = -6x + x$

❿ $-x + 2x/5 = -1$

⓫ $-3x - 1/3 = 2x/3 - x$

⓬ $-6 - x = 5x/3 - 6x/5$

⓭ $-4x/5 + 4/5 = 5x/2 + x$

⓮ $-3x + 3 = 1 + 5x/2$

⓯ $3x = x/3 + 3x/4$

⓰ $3x - 6x/5 = 2x + 4/3$

⓱ $x/5 + 3/2 = -x/5 - x/2$

⓲ $-5x/3 + 1 = 2x$

❶ $6/5 - 2x/3 = 3x/4 - 1/5$

❷ $-5x/6 = -x - 1/2$

❸ $-2x/3 + 3 = x/3 + 6x/5$

❹ $1/3 = -x/3 - 4x/3$

❺ $3x/5 - x = x/4 + 1$

❻ $x + 1/3 = 5/3 - 5x/6$

❼ $3/4 = -5x/4 - 3/2$

❽ $-2x/3 - x/2 = 4/5 - 3x/2$

❾ $-x/3 - x/5 = 4$

❿ $1/2 + x/2 = x/4 - x$

⓫ $-1/4 - x = -2x/3 - 3x$

⓬ $3 = -6x/5 + 3/2$

⓭ $5x = 2 + 6x$

⓮ $-3x/4 - x/4 = 2/3 - 3x/4$

⓯ $x/3 + 2x = -2/5 + 6x$

⓰ $-6x - 4 = -x$

⓱ $-x/5 - 4 = 1/4$

⓲ $-5x/2 + 5 = -2x + 5x/6$

❶ $3/5 = 6x + 3x$

❷ $-5/3 - x/6 = 5x/4$

❸ $x/2 - 5/4 = -x/5 - 2$

❹ $1 - 4x/3 = 3x/2 + 4/3$

❺ $-5x/2 = -3x/4$

❻ $-4x/5 - 3x/2 = 3/4$

❼ $-6/5 - 2x = 5x/3 + x$

❽ $2x - 2x/5 = -x/4$

❾ $x - x = x/3 - 1/2$

❿ $x/5 = -3x/5 - 2$

⓫ $6/5 = 3x/2$

⓬ $5/2 + 5x/6 = -x/3 - 3x$

⓭ $3/2 + x/2 = -x/2 + 3$

⓮ $4x/3 = -x/3 + 3/2$

⓯ $1 = 3/2 + 5x/2$

⓰ $-4x/5 + 2 = -5x/4 - 2/5$

⓱ $3x/2 = 5x/4 + 2x$

⓲ $-6x - 1 = -4x/3 + 1$

❶ $-5x/4 + 3/4 = -3x/5 + 3x$

❷ $-1/2 + 3x/2 = 3/2 - 3x/2$

❸ $x - 3/4 = 5x/4 - 6/5$

❹ $-3x/4 + x/2 = 1$

❺ $3x/2 = 2x - x/4$

❻ $-5/4 - x/2 = x/3 - 3x$

❼ $1 = x + 5x/4$

❽ $-x/2 + 4/3 = 3x/5 + 5/2$

❾ $-x/2 + x/3 = -1$

❿ $6x - 1/2 = 5 + x/5$

⓫ $1 - x = 3/2 + 4x/5$

⓬ $-4 - x/3 = -2x/3 - x/5$

⓭ $-x + x = -5/2 + x/2$

⓮ $-5/4 + x = x + 5x/6$

⓯ $-5 + 5x/4 = x/6 + x$

⓰ $5x/4 - 2x = 1 - 5x/4$

⓱ $1/2 + 5x/4 = 3x/5 - 2$

⓲ $-1/4 - 3x = -2x + 3x$

❶ $2x/5 + 2 = x/3 - 4/3$

❷ $3x/2 + 2 = 2x$

❸ $6/5 = 3x/2$

❹ $4x/3 + 2 = x/2$

❺ $3/2 + 5x/3 = 6/5 + 2x/3$

❻ $-3x/5 + 4 = -2x/3 - 1$

❼ $3x = -5x/6 - 3/2$

❽ $3/4 + x/6 = -2/5$

❾ $-2x/3 + 3/4 = -3 - x/2$

❿ $1 - 5x/2 = 3 - 2x$

⓫ $2/3 = -x - 5/3$

⓬ $3/2 + x/2 = -2x + x/6$

⓭ $2/5 = 3x/4 + x$

⓮ $2x/3 - 6/5 = -x + 4$

⓯ $-5x/3 + x = 1$

⓰ $-1 + 3x/4 = -x - 4x$

⓱ $-5/2 - x = -x - 5x/6$

⓲ $x/2 - 2 = 5x/4 + 5x/4$

❶ $x + 3 = -4x/3 - 2x/5$

❷ $6/5 = 2x/3 + 4/3$

❸ $-1/6 + 5x/6 = -3x - 2x$

❹ $x/3 = 3x/2 - 3/4$

❺ $3x/4 = 3x - 1/4$

❻ $-3 - 2x = 6x$

❼ $5/4 - 5x/2 = 3x/2 - 2$

❽ $3x - 6x = x/2$

❾ $-6/5 - 3x/5 = -2x + 2x/3$

❿ $1 - 5x/4 = 3x - 3/2$

⓫ $-x/2 - 1/6 = 2x + 3$

⓬ $x - 6x = x/5 + 1$

⓭ $-5/3 + 3x/5 = -1/5 - 2x$

⓮ $-3x + 3 = 5/3 + 5x/4$

⓯ $-2x + 2/3 = -x + 4x/3$

⓰ $-4 = -2x - x$

⓱ $x - 1/4 = -x/2 + x/2$

⓲ $-x - 3/5 = -3x/2 - 5x/2$

❶ $-3/2 - 2x = 5x/4 - 3/4$

❷ $-5/4 + 3x/2 = -6 - 3x/2$

❸ $6 - x/5 = 4/3 - 5x$

❹ $3 = -x + 6/5$

❺ $-2x/3 - 1 = 5/2 + 3x$

❻ $-3/4 + 5x/3 = -3/5 + 3x/2$

❼ $x + 1 = x/5 - x/2$

❽ $x/4 + x/2 = 3x/2$

❾ $-x/2 + x = 3x/2 - 2/3$

❿ $1 - 3x/2 = -3x/4 + x$

⓫ $-5/4 - x/2 = x$

⓬ $-x - 4x/3 = 4x - 6$

⓭ $5x + 4x = 1 + 5x/4$

⓮ $-2x = -1 - 4x$

⓯ $2x/3 - 4/5 = -2/3$

⓰ $-x - 1/5 = -6x/5$

⓱ $-6x + 2x/3 = 3x/2 - 3$

⓲ $3x/2 - 1 = 1 - x/3$

❶ $-x/3 = -x/4 + 1$

❷ $-5x/4 + 6x/5 = -5/4 + 3x/4$

❸ $-2/5 + 3x = -3/5 + 5x/4$

❹ $2/3 = 2x - x/3$

❺ $-x/4 - 3/2 = 2x + 5x/2$

❻ $-2 - 2x = 2 + 2x/5$

❼ $1 + x/5 = -3 + 2x$

❽ $-5x - 3/4 = 1 - 5x/6$

❾ $2x - 2x/3 = 3/5$

❿ $-x - x/3 = -5x/4 + 2$

⓫ $2 - x/2 = 3 - 5x$

⓬ $-3x/5 + 1/3 = x/3 - 4/3$

⓭ $2/3 - x/2 = -5/3$

⓮ $1/2 + 4x/5 = 6x/5$

⓯ $-1 + x/6 = -2/3 + 5x/3$

⓰ $-2x - 5/2 = 3x/5$

⓱ $-3/5 + 3x/5 = 1/6 - x/5$

⓲ $1/3 - x/2 = -x + 2$

❶ $3x/2 - 1 = x$

❷ $-x/2 = 4x + 3x/4$

❸ $-1/5 + x/2 = -1$

❹ $x + 2 = 4x/5$

❺ $-6 - 3x/2 = -2x/3$

❻ $-4/5 - 3x = -5 + 2x$

❼ $5x/3 - 6x/5 = -x + 3/4$

❽ $3x/2 = -5x/2 + x/2$

❾ $-x/2 + 1 = 5/2$

❿ $4x/5 = -1/3 - 3x$

⓫ $-3x + 5x/2 = 6/5$

⓬ $5x - x/3 = -6 - x/2$

⓭ $-4x/5 - 1 = 2/5$

⓮ $-2x + 1 = 3x/2$

⓯ $5x/4 - x = -3x - 1$

⓰ $3/5 - 3x/2 = -4x/5 - 2$

⓱ $4/5 + x/3 = 1/4$

⓲ $4 = 3/4 - 6x/5$

❶ $5x/4 + 3x/2 = 3x/2 - 1$

❷ $-x + 5x/2 = 4 - x/2$

❸ $1/2 + 2x = x/6 + 2x$

❹ $2/5 + 5x/2 = x/3 - 1/5$

❺ $2/3 + x = 6x/5 + 2/3$

❻ $-x/3 - 1 = -2x/3 + x/2$

❼ $5x/4 - x = 2 - 5x/2$

❽ $3x/2 = -2/5 - 5x/6$

❾ $-1/2 = -1 - x/2$

❿ $-2x = -3/5 + x/5$

⓫ $-4 - 2x = 1/4$

⓬ $-3x/2 + 2/5 = 5/6 - x/5$

⓭ $3x + 1 = 4/3$

⓮ $6x/5 + 1 = 5x/6$

⓯ $x - 2x = -3x/2 - 1$

⓰ $1 + 2x = x + 5/2$

⓱ $-1/6 + 2x = -x - 1/5$

⓲ $x + 4 = 3/5 - 3x$

❶ $5x/2 - 5/2 = -x - 4/3$

❷ $6x/5 + 6 = 2x + 4x/5$

❸ $-6x - x = 1/4 - 2x$

❹ $-2x + x/5 = -3x - 2/5$

❺ $-3/4 - x = 5x/2 - 2$

❻ $x - 6 = 5x/2 - 2/5$

❼ $2x/3 - 2x = -5/3 + 4x/3$

❽ $-x/5 + 6 = 1/3$

❾ $x - 5 = 2x + 3x/2$

❿ $1 + 2x/3 = 3$

⓫ $-x + 5/3 = 5x + 1$

⓬ $-3x/2 + 3x/2 = -x/4 - 2/3$

⓭ $2 - x/3 = 4x$

⓮ $-6x/5 + 2 = x/3 + 5x/6$

⓯ $-2x/3 - 3x/5 = 3x/2 + 2/3$

⓰ $-2x - 1/6 = 3/2 - x/3$

⓱ $1/4 - 3x = 2/3$

⓲ $-5x/3 - 1 = -3x/2 + x/4$

❶ $2x/5 + 3/5 = -x$

❷ $-2/3 = 3 + 2x/3$

❸ $-2/3 - x/4 = -1/6$

❹ $2x - x = 2x/5 - 3/2$

❺ $-1 - 3x/2 = -4x/3 + 3x/2$

❻ $-5x/6 - 2/3 = 2x + 5/3$

❼ $5x = -x + x$

❽ $5/3 - 5x/2 = x + 1/2$

❾ $4x/5 = -4x/3 - x$

❿ $1/6 = 2x/3$

⓫ $1 + x/2 = -x - 2x/3$

⓬ $-2x = -2x/5 + 1/3$

⓭ $-x/6 + 1 = 6x/5 - 3/5$

⓮ $5x - 1 = -2x/3 - 6$

⓯ $3x/2 + 3 = 4/5 + 6x$

⓰ $1/2 + 3x = x/2 - 2$

⓱ $1/4 - 5x/3 = -6 + x/6$

⓲ $x/2 - 6 = 3x/4$

❶ $-3x/4 - x = -1/2$

❷ $x/2 - 3 = -1/3 - 4x$

❸ $-5x/4 - 3x = 3/2 - 5x/6$

❹ $-3/2 - 4x/3 = -6x + 1/4$

❺ $4 - 3x/5 = 1/2$

❻ $-3/2 + 4x/3 = -x/2 - 3x/4$

❼ $6x - 2x = -2/5 - 4x/5$

❽ $5/4 - x/3 = 6x/5$

❾ $5/4 + 2x/3 = -2/3$

❿ $2 + x/3 = -1 - 3x$

⓫ $-4 + 5x = x/3 - 6/5$

⓬ $3x/2 + 3/4 = 1 + 3x$

⓭ $-4x/3 + 1/4 = -x/4 - x/5$

⓮ $-2x/3 - 5x/2 = x - 2$

⓯ $2 - x/2 = x + 4x/3$

⓰ $1/3 = 4/5 - 3x$

⓱ $6x + x/6 = 5x/4 + 1$

⓲ $x/3 - 1/4 = x + x$

❶ $-6 - x/2 = x/3$

❷ $-3x/2 + 1/2 = -1/5 - x/6$

❸ $-1/4 - 2x/3 = 1/6 + x$

❹ $3 - 4x/3 = -x/6 - 6$

❺ $1/2 = 3x/4 - 2/5$

❻ $3x/2 + x = 2 - 3x$

❼ $2x/3 - 5/6 = 3x/4 + 1$

❽ $5/2 = 1 - x/5$

❾ $2 + x/2 = -2x/3 - 2$

❿ $-4x + 3/2 = 6/5 - x/5$

⓫ $2x + 2 = -2x + 2x$

⓬ $-2/5 - x = 2x/3 + x/3$

⓭ $1 = -x/2 - x/2$

⓮ $-4x - 1/3 = 3x/2 + x$

⓯ $-4x/5 - 1/5 = -x/3$

⓰ $-3x/5 + 1/3 = 4x - 3x/2$

⓱ $5x/3 + 5/6 = -6/5$

⓲ $-4x/5 + 3/2 = -1/2 + 4x/5$

❶ $-1 = -2/5 - 2x/3$

❷ $-5x + 5 = 2x/5 + 2$

❸ $-1/2 + 2x = -4x/5 - 3x$

❹ $x/2 - 2/5 = -5x/6$

❺ $x/2 - 5x/4 = 4x/3 + 5/6$

❻ $1 - x = -5x/6 - 1$

❼ $1 = x + 3/2$

❽ $5/6 + 2x/3 = 3x - 6x$

❾ $3 - x = -x/6 + x/2$

❿ $-4 = -x + 4x/3$

⓫ $-2x/5 + 3 = 3x/5 + 5/3$

⓬ $2x = 2x/5 - 5x$

⓭ $3x/4 = -1/2 - 4x/3$

⓮ $2x/3 + x = -4x/5$

⓯ $-x + 4x = x/2 - 5$

⓰ $-1/6 + x/6 = 1/2 + 5x/2$

⓱ $-1 - 2x = -2 + 4x/3$

⓲ $1 + 3x/4 = 2x - 2/5$

❶ $2x/3 - x/3 = -1/2 + 5x/6$

❷ $-1 - 6x/5 = -1/3 - x$

❸ $5x/6 + 6/5 = 1/5 + 3x/4$

❹ $-x - 3/4 = -1/3 + x$

❺ $-4x - 1/3 = x/3 + 6$

❻ $2 + x/3 = -3x/2 + 4x/3$

❼ $-4 + 4x/5 = -x$

❽ $-2x/3 - 5x/4 = 5$

❾ $-1/3 + 2x = -5x/4 + 3/2$

❿ $2x/5 - 1/2 = 1/5$

⓫ $2/3 + 3x/2 = -2/5 + 5x/3$

⓬ $-2x + 1 = 3/4 + 2x/3$

⓭ $2/3 + 2x/5 = 3/2 + 2x/3$

⓮ $4x/3 + 4x = x/4 + 1$

⓯ $1/6 - x/4 = -1/3 - 3x/4$

⓰ $-x/3 + 5/3 = -2x - 5/2$

⓱ $5x/3 + 3 = 3/4$

⓲ $-3 - 2x/5 = 2/3$

Chapter 3: Simple Quadratic Equations

Following are a couple more definitions that will be very useful for Chapters 3 thru 5.

- A **quadratic** equation consists of terms that are proportional to the square of the variable, and may also include terms that are linearly proportional to the variable and constant terms. A general quadratic equation can be cast in the form $ax^2 + bx + c = 0$, where a, b, and c are constants. As examples, $4x^2 - 2x = 12$, $2x^2 = 5 - 9x$, and $3x^2 = 27$ are quadratic equations. Contrast this to linear equations (which we learned in Chapter 1), which can always be condensed to the form $ax + b = 0$ (for example, the equation $3x + 4 = 7x - 2$ can be simplified to $4x - 6 = 0$).
- When two numbers are multiplied together, the numbers being multiplied are called **factors** and the result is called the **product**. For example, in $(5)(8) = 40$, the 5 and 8 are called factors and the 40 is called the product. The prime factors of 40 are $(5)(2)(2)(2)$, since the 8 factors into $(2)(2)(2)$. (A prime number is only evenly divisible by itself and 1.) You can also have algebraic factors. For example, $3(x + 2) = 3x + (3)(2) = 3x + 6$. In this case, the factors are 3 and $x + 2$ and the product is $3x + 6$.

This chapter is dedicated to simple quadratic equations, where either the linear or constant term is not present. For example, the equation $5x^2 - 3 = 12$ has no linear term (like 2x) and the equation $3x^2 + 2x = 8x$ has no constant term. These quadratic equations are much easier to solve than the general quadratic equation (which is the subject of Chapters 4 and 5). We need to consider the basics of factoring here in Chapter 3, and will apply the technique of factoring extensively in Chapter 4. So for now we will discuss the basics of factoring, and then we will describe factoring in more detail in the next chapter.

In the introduction to Chapter 2, we briefly discussed the principle of determining multiplicative factors of numbers in the context of fractions. We used this idea to find lowest common denominators (to add or subtract fractions) and greatest common factors (to reduce fractions). For example, 42 can be factored into 6 and 7: $(6)(7) = 42$. It can also be factored into 2 and 21: $(2)(21) = 42$. The prime factors of 42 are 2, 3, and 7: $(2)(3)(7) = 42$.

In algebra, the concept of factoring is related to the distributive property: $a(b + c) = ab + ac$. Here, a and $(b + c)$ are factors, and in the product, $ab + ac$, the factor a is said to have been distributed to b and c. If you try this with numbers, you will see that it works. For example, $3(4 + 2) = 3(6) = 18$, which agrees with $3(4) + 3(2) = 12 + 6 = 18$. That is, $3(4 + 2) = 3(4) + 3(2)$. The distributive property also applies to algebra. Here are a few examples: $2(x + 3) = 2x + 2(3) = 2x + 6$, $(x + 2)(4) = 4x + 2(4) = 4x + 8$ and $4x(3 - x) = 12x - 4x^2$. Note that minus signs also distribute. For example, $-3(x + 4) = -3x - 3(4) = -3x - 12$ and $-(x - 1) = -x - (-1) = -x + 1$.

Factoring algebraic expressions is essentially the opposite of distributing. To factor an algebraic expression, such as $6x + 9$, think about what multiplicative factor each term has in common. In the case of $6x + 9$, the greatest common factor is 3. The 3 can be factored out as $3(2x + 3)$. By the distributive property, we see that $3(2x + 3) = 6x + 9$. As another example of factoring, consider the expression $4x^2 - 6x$. This time the greatest common factor is $2x$. We need to multiply $2x$ by $2x$ to get $4x^2$, and we need to multiply $2x$ by -3 to make $-6x$. Therefore, $4x^2 - 6x$ can be factored as $2x(2x - 3)$.

Following are a few examples of factoring simple algebraic expressions. In Chapter 4, we will learn how to factor the general quadratic equation.

Example 1: Factor $15x + 9$.

3 is common to $15x$ and 9.
$15x$ divided by 3 equals $5x$.
9 divided by 3 equals 3.
$15x + 9 = 3(5x + 3)$.

Example 2: Factor $8x^2 - 4x$.

$4x$ is common to $8x^2$ and $-4x$.
$8x^2$ divided by $4x$ equals $2x$.
$-4x$ divided by $4x$ equals -1.
$8x^2 - 4x = 4x(2x - 1)$.

Example 3: Factor $3x^2 + 12$.

3 is common to $3x^2$ and 12.
$3x^2$ divided by 3 equals x^2.
12 divided by 3 equals 4.
$3x^2 + 12 = 3(x^2 + 4)$.

Example 4: Factor $-7x - x^2$.

$-x$ is common to $-7x$ and $-x^2$.
$-7x$ divided by $-x$ equals 7.
$-x^2$ divided by $-x$ equals x.
$-7x - x^2 = -x(7 + x)$.

The difference between the algebraic strategies of this chapter and the previous chapters has to do with the x^2 terms. There are two distinct types of equations in this chapter: One type consists of quadratic (x^2) and linear (x) terms, like $4x^2 - 3x = 6x$, and the other type consists of quadratic (x^2) and constant terms, such as $2x^2 - 9 = 7$. The first type is solved by factoring, while the second type has a solution similar to the strategy of Chapter 1. In general, there are two possible solutions to quadratic equations. For example, consider the equation $x^2 = 4$. One possible solution is $x = 2$ because $(2)^2 = 4$. However, $x = 2$ is not the only solution. Observe that $x = -2$ also solves the equation: $(-2)^2 = 4$. The complete solution to the equation $x^2 = 4$ is $x = -2$ or 2. It is necessary to state both answers because both solve the equation.

Notice that the simple equation $x^2 = 4$ is solved by taking the squareroot of both sides of the equation. Just like adding the same number to both sides of an equation or multiplying both sides of an equation by the same number, taking the squareroot of both sides of an equation preserves the equality of the equation (this is the opposite of the squaring that is originally being done to the x). However, an important difference with the squareroot is that there are generally two solutions possible. The solution may also be irrational. For example, the solutions to $x^2 = 3$ are $x = \pm\sqrt{3}$.

Some solutions to quadratic equations involve irrational numbers. Let's review some basics about irrational numbers before describing the algebraic strategies for this chapter. The distinction between rational and irrational numbers arises when we take the squareroot of a number. Squareroots result in rational numbers when the number you are squarerooting is a perfect square. For example, $\sqrt{16} = \pm 4$ and $\sqrt{25/9} = \pm 5/3$. Otherwise, they result in irrational numbers like $\sqrt{5}$.

The square of a number of the form $\sqrt{a}$ removes the squareroot. That is, $(\sqrt{a})^2 = a$. For example, $(\sqrt{7})^2 = 7$. Note that $(-\sqrt{7})^2 = (\sqrt{7})^2 = 7$, which is why $x^2 = 7$ has two solutions: $x = \pm\sqrt{7}$. Squareroots can be multiplied and factored. For example, $\sqrt{2}\sqrt{7} = \sqrt{2 \cdot 7} = \sqrt{14}$ and $\sqrt{15} = \sqrt{5 \cdot 3} = \sqrt{5}\sqrt{3}$. Similarly, $\sqrt{2/3} = \sqrt{2}/\sqrt{3}$.

It is good form (and often required by teachers) to factor out any perfect squares and to rationalize the denominator when the answer is irrational. A perfect square is a number for which the squareroot is rational, such as 36 or 81 (which equal 6^2 and 9^2). When the answer involves a squareroot, see if you can factor out any perfect squares. For example, consider $\sqrt{12}$. Since 12 can be factored as (4)(3) and 4 equals 2 squared, we can write $\sqrt{12} = \sqrt{4 \cdot 3} = \sqrt{4}\sqrt{3} = 2\sqrt{3}$. If the answer involves the squareroot of a fraction, rationalize the denominator as follows: Multiply both the numerator and denominator by the denominator (which equates to multiplying by one). For example, we can rationalize the denominator of $\sqrt{3/2}$ by multiplying the numerator and denominator both by $\sqrt{2}$: Then $\sqrt{3/2} = (\sqrt{3}\sqrt{2})/(\sqrt{2}\sqrt{2})$, which simplifies to $\sqrt{3/2} = \sqrt{6}/2$.

Here are some examples of rationalizing the denominator and factoring out perfect squares:

Example 1: $\sqrt{45} = \sqrt{9 \cdot 5} = \sqrt{9}\sqrt{5} = 3\sqrt{5}$.

Example 2: $\sqrt{\frac{1}{3}} = \frac{\sqrt{1}}{\sqrt{3}} = \frac{\sqrt{1}\sqrt{3}}{\sqrt{3}\sqrt{3}} = \frac{\sqrt{3}}{3}$.

Example 3: $\sqrt{\frac{80}{7}} = \frac{\sqrt{80}}{\sqrt{7}} = \frac{\sqrt{5}\sqrt{16}}{\sqrt{7}} = \frac{4\sqrt{5}}{\sqrt{7}} = \frac{4\sqrt{5}\sqrt{7}}{\sqrt{7}\sqrt{7}} = \frac{4\sqrt{35}}{7}$.

Following are the two strategies for solving the two types of exercises in this chapter.

Equations like $2x^2 - 9 = 4x^2 + 7$
(no linear terms, like 5x)

- Choose which side of the equation will correspond to variables and constant terms.
- Move constant terms over to the desired side: Subtract a positive constant from both sides, but add a negative constant to both sides. For example, $6x - 2 = 1$ becomes $6x = 1 + 2$.
- Move variable terms over to the other side using the same technique. For example, $5x^2 = 2x^2 + 9$ becomes $5x^2 - 2x^2 = 9$.
- Collect like terms together on each side of the equation. For example, $9x^2 - 4x^2 = 3 + 2$ simplifies to $5x^2 = 5$.
- Divide both sides of the equation by the coefficient of the quadratic term. For example, $2x^2 = 8$ becomes $x^2 = 4$.
- Take the squareroot of both sides of the equation. There are two possible answers: One answer is positive and the other is negative. For example, the two solutions to $x^2 = 9$ are $x = -3$ and $x = 3$, which can be expressed as $x = \pm 3$.
- Factor out any perfect squares and rationalize the denominator, if necessary. For example, $x = \sqrt{5/12}$ becomes $x = \sqrt{15}/6$.

Equations like $4x^2 - 3x = 5x^2 + 6x$
(no constant terms)

- There are no constant terms. Group all of the terms on the same side of the equation.
- Move terms over to the desired side using the same technique that we have previously applied: Subtract a positive term from both sides, but add a negative term to both sides. For example, $6x^2 - 3x = 5x$ becomes $6x^2 - 3x - 5x = 0$. One side of the equation will equal zero.
- Collect quadratic terms together and separately collect linear terms together. For example, $8x^2 + 9x - 5x^2 - 3x = 0$ simplifies to $3x^2 + 6x = 0$.
- Factor the variable side of the equation. For example, $6x^2 - 4x = 0$ can be factored as $2x(3x - 2) = 0$.
- One factor must be zero in order for the product to be zero. Set each factor equal to zero and solve for x using the strategy of Chapter 1. For example, for $5x(2x - 8) = 0$, the two equations are $5x = 0$ and $2x - 8 = 0$. The first solution is $x = 0$ and the second equation can be written as $2x = 8$, for which $x = 4$. The complete solution is $x = 0$ or 4.

Here is a note on the factoring technique. For an equation like $4x(2x - 6) = 0$, where two factors are multiplied together and the product equals zero, the only way for the product to equal zero is if one of the two factors equals zero. (That is, the only way to multiply two numbers and get an answer of zero is if one of the numbers being multiplied is zero.) Therefore, either $4x = 0$ or $2x - 6 = 0$. If $4x = 0$, then x must be zero; if instead $2x - 6 = 0$, then $2x = 6$ or $x = 3$. The two solutions, $x = 0$ and $x = 3$, both solve the equation.

Example 1: $4x^2 - 5 = x^2 + 7$. All of the constant terms will be on the right side if we move the 5 to the right. Since the 5 is negative, we add 5 to both sides of the equation: $4x^2 = x^2 + 12$. Now we subtract x^2 from both sides to get all of the variable terms on the left: $3x^2 = 12$. Divide both sides by the coefficient of x^2, which is 3: $x^2 = 4$. Squareroot both sides of the equation: $x = \pm 2$.

Example 2: $2x^2 - 3x = 5x^2 + 6x$. Let's move all of the variable terms to the right side. Since 3x is negative, we add 3x to both sides of the equation, and since $2x^2$ is positive, we subtract $2x^2$ from both sides: $0 = 5x^2 + 6x - 2x^2 + 3x = 3x^2 + 9x$. Next, we factor $3x^2 + 9x$. The greatest common factor is 3x. Since $3x^2$ divided by 3x equals x and 9x divided by 3x equals 3, it factors as: $3x(x + 3) = 0$. Either $3x = 0$ or $x + 3 = 0$. For the case $3x = 0$, we divide both sides by 3 to obtain $x = 0$. For the case $x + 3 = 0$, we subtract 3 from both sides to find that $x = -3$. The two solutions are $x = -3$ and $x = 0$.

Example 3: $6x^2 - 9 = 3 + 4x^2$.

$$6x^2 = 12 + 4x^2$$
$$2x^2 = 12$$
$$x^2 = 6$$
$$\boxed{x = \pm\sqrt{6}}$$

Check: $6x^2 - 9 = 6(\pm\sqrt{6})^2 - 9 = 36 - 9 = 27$.
$3 + 4x^2 = 3 + 4(\pm\sqrt{6})^2 = 3 + 24 = 27$. ✓

Example 4: $x^2 - 8 = -4x^2$.

$$x^2 = -4x^2 + 8$$
$$5x^2 = 8$$
$$x^2 = 8/5$$
$$x = \pm\sqrt{8/5} = \pm\sqrt{8}/\sqrt{5}$$
$$x = \pm(\sqrt{8}\sqrt{5})/(\sqrt{5}\sqrt{5})$$
$$x = \pm\sqrt{40}/5 = \pm\sqrt{4 \cdot 10}/5$$
$$\boxed{x = \pm 2\sqrt{10}/5}$$

Check: $x^2 - 8 = (\pm 2\sqrt{10}/5)^2 - 8$
$= 40/25 - 8 = 8/5 - 8 = -32/5$.
$-4x^2 = -4(\pm 2\sqrt{10}/5)^2 = -160/25 = -32/5$. ✓

Example 5: $5x^2 - 7x = -4x + 3x^2$.

$$5x^2 - 3x = 3x^2$$
$$2x^2 - 3x = 0$$
$$x(2x - 3) = 0$$
$$x = 0 \quad \text{or} \quad 2x - 3 = 0$$
$$x = 0 \quad \text{or} \quad 2x = 3$$
$$x = 0 \quad \text{or} \quad x = 3/2$$
$$\boxed{x = 0, 3/2}$$

Check: $5(0)^2 - 7(0) = -4(0)^2 + 3(0) = 0$. ✓
$5(3/2)^2 - 7(3/2) = 45/4 - 21/2 = 3/4$.
$-4(3/2) + 3(3/2)^2 = -6 + 27/4 = 3/4$. ✓

Example 6: $8x^2 + 3x = 4x^2 - x$.

$$8x^2 + 4x = 4x^2$$
$$4x^2 + 4x = 0$$
$$4x(x + 1) = 0$$
$$4x = 0 \quad \text{or} \quad x + 1 = 0$$
$$x = 0 \quad \text{or} \quad x = -1$$
$$\boxed{x = -1, 0}$$

Check: $8(0)^2 + 3(0) = 4(0)^2 + 3(0) = 0$. ✓
$8(-1)^2 + 3(-1) = 8 - 3 = 5$.
$4(-1)^2 - (-1) = 4 + 1 = 5$. ✓

Here are a couple more notes:

- It may be tempting to divide both sides by x in an equation like $3x^2 - 6x = 0$. However, one solution is $x = 0$, and you better not divide by zero. The correct method is to factor.
- In general (but not in this workbook), not all quadratic equations have real solutions. Imaginary solutions result from the squareroot of a negative number, as in $x^2 = -4$.

❶ $8x^2 = 6x^2 + 5x^2$

❷ $-7x^2 + 6 = x^2 - 7x^2$

❸ $7x^2 + 9x^2 = -6x^2 - 7x$

❹ $2x^2 - 5 = -7 + 4x^2$

❺ $-4x^2 + 6x = -8x - 3x^2$

❻ $-2x^2 + 8x = 6x^2 - 4x^2$

❼ $2x^2 + 8x = 9x^2 - 6x^2$

❽ $-3x + x^2 = 4x$

❾ $x^2 + 9x^2 = 5x^2 + 7$

❿ $-9x^2 = -9x^2 + 2x^2$

⓫ $3x = -8x^2 - 9x$

⓬ $4x + 5x^2 = 7x^2 - 8x^2$

⓭ $5 - 6x^2 = -3x^2 - 9$

⓮ $8x^2 + 2x = -2x - 7x^2$

⓯ $3x^2 + x = -7x^2 + 6x$

⓰ $9x^2 = x - 9x^2$

⓱ $6x^2 - 9x = -8x^2$

⓲ $-6x^2 + 2x^2 = -6 - 2x^2$

❶ $2x^2 + 8x^2 = 7x + 6x^2$

❷ $-8x^2 + 1 = -1 + 3x^2$

❸ $5x^2 - 7 = -6$

❹ $-4x^2 + 7x^2 = 4x^2 - 8x$

❺ $-4 - 9x^2 = -5 + 7x^2$

❻ $-2x^2 + 9x^2 = -8x^2$

❼ $3x^2 + 2x^2 = 3$

❽ $8x - 7x^2 = 2x + 4x^2$

❾ $-5x^2 = -x + 5x^2$

❿ $4x^2 + 6x^2 = -3x^2 + 4x$

⓫ $-9x + 6x^2 = 2x$

⓬ $-5x^2 - 3x = -8x^2 + 6x^2$

⓭ $-2x^2 - 6x^2 = -8x$

⓮ $x^2 = 2x^2 - 7$

⓯ $4x^2 - 5x^2 = -8x^2$

⓰ $-7x + 3x^2 = x^2 + 3x^2$

⓱ $-7x^2 + 5x = 2x^2 - x$

⓲ $-3x - 2x^2 = 9x - 5x^2$

❶ $-8x - 7x^2 = -5x^2 + 6x$

❷ $-9x^2 - 8x^2 = -8x^2$

❸ $-6 + 6x^2 = x^2 + x^2$

❹ $-8x^2 + x = -3x^2 + 4x$

❺ $2x^2 = x^2 + 2x^2$

❻ $7 - 9x^2 = -6x^2 - 2x^2$

❼ $-6x - 3x^2 = -9x^2 + 5x^2$

❽ $-3x^2 - 2x^2 = 6x^2 + 7x$

❾ $7x^2 + 6x^2 = -9x - 8x^2$

❿ $9x^2 = 7x^2$

⓫ $-5x + 8x^2 = -6x^2 + 7x$

⓬ $-2x^2 = 2 - 8x^2$

⓭ $-7x + 8x^2 = -8x^2 + 9x^2$

⓮ $3x + 2x^2 = 8x^2 + 7x$

⓯ $-7x^2 + 8 = -x^2 - 2x^2$

⓰ $4x^2 - 8x^2 = 2x^2 - 6$

⓱ $-8x^2 + 4x = 4x$

⓲ $-9x^2 + x^2 = 9x + x^2$

1. $6x^2 + 9x = 4x + 8x^2$

2. $5x^2 - 7 = -5x^2 - 3$

3. $9x^2 + 8x = 3x^2 + 4x$

4. $5x - 9x^2 = 3x - 4x^2$

5. $-x^2 - 9x^2 = x + 4x^2$

6. $5x = 7x^2 - 8x^2$

7. $9x^2 + 7x^2 = 7x^2$

8. $7x^2 + 6x^2 = -2x^2 + 7x$

9. $2x^2 = -8x^2 + 4x^2$

10. $4x^2 + 2x^2 = -9x^2 - x$

11. $-x^2 + 7x = -2x^2 + 6x$

12. $7x^2 - 4x = 6x^2 - 8x^2$

13. $-8x + 9x^2 = -6x^2 + 5x$

14. $-5 - 4x^2 = -3x^2 - 2x^2$

15. $-2x^2 + 2x^2 = 7x^2$

16. $-9x^2 - 6x = 3x^2 + 7x$

17. $-7x^2 + 9x^2 = 3x^2$

18. $-8x^2 + 5x = -5x - x^2$

❶ $7 + 5x^2 = 3x^2 + 4x^2$

❷ $7x^2 + 9x^2 = -x^2 - 2x$

❸ $5x^2 = -6x - 4x^2$

❹ $-2x - 6x^2 = 3x^2 + 7x^2$

❺ $3x = -7x^2 - 4x^2$

❻ $-6x^2 + 7 = x^2 + 2x^2$

❼ $6x^2 + 4x^2 = -5x - 3x^2$

❽ $3x^2 + x = -8x - 6x^2$

❾ $4x^2 - 5x^2 = 4x^2$

❿ $-2x + 5x^2 = -6x - 5x^2$

⓫ $8x^2 + 4x = -4x - x^2$

⓬ $2 + 4x^2 = -5x^2 + 4$

⓭ $-5x^2 - 8 = -6x^2 - 5$

⓮ $9x - 4x^2 = 9x + x^2$

⓯ $-7x - 3x^2 = -5x + 4x^2$

⓰ $x^2 - 4 = 2$

⓱ $5x^2 - 2x = 9x + 9x^2$

⓲ $-4x^2 + 3 = -9 + 9x^2$

1. $-x^2 + 5 = -6\,x^2 + 5$

2. $9\,x^2 - 1 = 2\,x^2 + 3$

3. $-2 + 8\,x^2 = -9\,x^2 + 4$

4. $-x^2 + x^2 = -6 + 9\,x^2$

5. $5\,x^2 + 4\,x^2 = -3\,x^2 + 8\,x$

6. $2\,x + x^2 = 8\,x + 5\,x^2$

7. $-9\,x^2 + 6\,x = -7\,x^2 - 9\,x$

8. $-6\,x^2 - 2\,x^2 = -4\,x^2 - 9$

9. $3 + 5\,x^2 = -x^2 + 7\,x^2$

10. $5\,x = 2\,x^2 + 6\,x$

11. $-6\,x^2 - 9\,x^2 = x^2 + 9\,x$

12. $7\,x^2 + 8\,x^2 = 5 - x^2$

13. $9\,x^2 - 8\,x = 3\,x^2$

14. $-3\,x^2 + 5\,x^2 = 7\,x^2 - 4$

15. $2\,x = 9\,x^2 + 4\,x$

16. $x = -x + 3\,x^2$

17. $3\,x^2 + 6\,x^2 = x^2 - 9\,x$

18. $5\,x^2 - 3\,x^2 = -2\,x^2 + 8$

❶ $-3x^2 + 2 = 2 + 8x^2$

❷ $4x^2 + 8x^2 = 8x + 9x^2$

❸ $-6x^2 - 5x^2 = -3 + x^2$

❹ $-5x^2 + 3x^2 = -7 + 3x^2$

❺ $-7x^2 + 9x = x + 4x^2$

❻ $-6x - 9x^2 = -6x^2 - x$

❼ $-8x - 2x^2 = -9x + x^2$

❽ $-2 - 5x^2 = -7$

❾ $-8x^2 - x = 5x$

❿ $-8x^2 + 5 = 4x^2$

⓫ $-4x^2 + x = -2x^2 - 3x$

⓬ $x^2 - 9x^2 = x + 7x^2$

⓭ $-5x^2 + x^2 = 6x$

⓮ $7x^2 - 6x^2 = 8x^2$

⓯ $-8x^2 - 7x^2 = -2x^2$

⓰ $-4x^2 - 2x^2 = -5x^2$

⓱ $-x + 9x^2 = 9x^2 + 8x^2$

⓲ $-8x - 2x^2 = -9x^2 - 2x^2$

❶ $-3x^2 + 8x^2 = 3x - 8x^2$

❷ $-8x = 9x^2 - 4x^2$

❸ $9x^2 - 8x = x^2 + 2x^2$

❹ $2x^2 - 9x = 9x^2 - 4x$

❺ $-8x^2 - 4x^2 = 3x + 9x^2$

❻ $-6x + 4x^2 = 2x$

❼ $-5x^2 + 2x = -7x^2$

❽ $9x - 5x^2 = 3x - 8x^2$

❾ $2x^2 - x = 5x^2 - x$

❿ $-2x - 8x^2 = 4x^2$

⓫ $-5x - 6x^2 = 7x^2 - x^2$

⓬ $9x^2 + 6x = 5x + 5x^2$

⓭ $-6x^2 + 9x = 4x^2$

⓮ $-3x^2 + 2x^2 = -6x - 5x^2$

⓯ $8x^2 - 5x^2 = -7x^2 - x$

⓰ $1 - 5x^2 = 7x^2 - 3x^2$

⓱ $2x^2 + 2x^2 = 2 - 5x^2$

⓲ $6x + 7x^2 = -6x^2$

❶ $-7x^2 + 7x = 7x - x^2$

❷ $-6x^2 - 9x^2 = 3x^2$

❸ $6x^2 - 8x = 9x^2 - 8x$

❹ $8x^2 + 5x^2 = 7x^2 + 2x$

❺ $-8x^2 - 7x^2 = -8 + 6x^2$

❻ $x + x^2 = -7x^2 - 8x^2$

❼ $-5x^2 + 4x = -8x^2 - 6x^2$

❽ $6x + 5x^2 = 9x + 8x^2$

❾ $-5x + 3x^2 = 5x - x^2$

❿ $-6x^2 - 5x^2 = 9x^2 - x$

⓫ $9x = 9x^2 - 3x^2$

⓬ $2x = 7x^2 - 5x$

⓭ $-4x - 6x^2 = -9x^2 + 4x$

⓮ $8x^2 - 3 = -3 - 9x^2$

⓯ $4x^2 - 6x = -3x^2 - 2x$

⓰ $-4x^2 + 9 = 6x^2 - 8x^2$

⓱ $2x^2 + 3x^2 = 2x - 7x^2$

⓲ $x^2 = -5x^2 + 6$

❶ $-2 + 7x^2 = -7 + 9x^2$

❷ $8x - 7x^2 = 9x^2 + 2x$

❸ $-8x^2 - x = 3x - 4x^2$

❹ $-9x^2 + 9x^2 = -7 + x^2$

❺ $7x^2 + 4x = 6x^2 + 8x$

❻ $9x^2 + 4 = 9 + 2x^2$

❼ $-x + 4x^2 = 6x - x^2$

❽ $-3x^2 - 6x^2 = -4x^2 - 2x$

❾ $-4x^2 + 6x^2 = x^2 + 4$

❿ $x^2 - 6x = -x^2 + 6x^2$

⓫ $-9x^2 - 3x = -4x + 2x^2$

⓬ $7x^2 - x = 3x - x^2$

⓭ $-8 + 6x^2 = -5x^2 + x^2$

⓮ $-x^2 + 8 = -5$

⓯ $-1 + 2x^2 = 4$

⓰ $6x + 8x^2 = -6x^2 - 4x$

⓱ $6x = -3x^2$

⓲ $9x + 6x^2 = x^2 - 2x^2$

❶ $2x + 4x^2 = 5x^2 - 6x^2$

❷ $-x^2 - 5 = 2 - 8x^2$

❸ $9x^2 + 8x = -2x - 9x^2$

❹ $-8x^2 - 5 = -4x^2 - 5x^2$

❺ $5 + 6x^2 = 7x^2 + 7x^2$

❻ $-2x = 6x^2 + 2x$

❼ $9 = -2 + 4x^2$

❽ $8x^2 - x = -3x^2 - 8x$

❾ $4x^2 - 6 = 3x^2 - 7x^2$

❿ $9x^2 - 7x^2 = -9x + 7x^2$

⓫ $7x^2 - 9x = 2x^2 + 6x^2$

⓬ $9x + 7x^2 = 8x^2 - 7x$

⓭ $-3x^2 + 4x^2 = 6x^2 + 4x$

⓮ $-7x^2 + 8x^2 = -3 + 5x^2$

⓯ $3x^2 + 6x^2 = 3 - 9x^2$

⓰ $-x - 5x^2 = 2x^2 + 6x^2$

⓱ $-x - 6x^2 = -x^2 - 9x$

⓲ $2x^2 - 3 = -x^2 - 9x^2$

❶ $7x^2 = -8x + 6x^2$

❷ $-x^2 + 8x^2 = 2x^2 + 3$

❸ $-4x^2 - 2 = -2 + 5x^2$

❹ $-6x - 5x^2 = -8x$

❺ $9x^2 - 3x^2 = -3x^2 - 5x$

❻ $5x^2 + 6x = 8x^2 + 8x$

❼ $-3x^2 + 8 = -9x^2 + 7x^2$

❽ $-2x^2 + 5x^2 = 4x - 3x^2$

❾ $-3x^2 + 5x = -2x^2 + 9x$

❿ $4x^2 = 9x + x^2$

⓫ $7 + 7x^2 = 6x^2 + 5x^2$

⓬ $-7 = 4x^2 - 7x^2$

⓭ $x^2 - 6x^2 = -7x + 8x^2$

⓮ $6x^2 + 6x = 8x + 3x^2$

⓯ $-8 - 2x^2 = -8x^2 + 4x^2$

⓰ $-3x^2 + 5x = 3x^2 + 4x^2$

⓱ $8x + 9x^2 = 4x^2 + 9x^2$

⓲ $2x^2 = -x^2 + 4x^2$

❶ $-5x^2 + 5x^2 = 5x^2 - 8x$

❷ $-x^2 + 2x^2 = -x + 4x^2$

❸ $-6 - 2x^2 = -7x^2 + 9$

❹ $2x^2 - 8 = -8 + 5x^2$

❺ $-3x - 7x^2 = 5x^2 - x^2$

❻ $-9x - x^2 = 4x^2 - x^2$

❼ $-5x^2 - 3x^2 = -9x^2 + 4x$

❽ $-2x^2 - 3x = 2x^2 - 7x^2$

❾ $9x + 5x^2 = 3x^2 - 2x^2$

❿ $4x^2 + 6x = 2x^2 + 7x^2$

⓫ $6x^2 + 2x = -5x^2 - x^2$

⓬ $4x + 3x^2 = 8x^2 + 6x^2$

⓭ $-9x^2 + 4 = 3x^2$

⓮ $5x^2 + 7x^2 = -6x^2 + 3x$

⓯ $-6x + 5x^2 = 8x^2 - 4x^2$

⓰ $2 + x^2 = 4x^2 + 1$

⓱ $5x^2 - 4x = 4x - x^2$

⓲ $7x^2 - 4x = 2x^2 + 9x$

❶ $-x - 7x^2 = -9x^2 + 4x$

❷ $2x^2 = 8x$

❸ $-4x^2 = -6x^2$

❹ $3x^2 = -9x^2 + 5$

❺ $9x^2 + 5 = 7x^2 + 8x^2$

❻ $-1 - 2x^2 = -3x^2$

❼ $2x^2 - 6 = -8x^2 + 5x^2$

❽ $-3 - 5x^2 = 8x^2 - 3$

❾ $-5x + 7x^2 = 3x^2 + 7x^2$

❿ $-3x^2 - 5x^2 = -3x - x^2$

⓫ $-6x - 5x^2 = x^2 - 8x^2$

⓬ $-6 = -9x^2 - 7x^2$

⓭ $-3x^2 + 5x = 6x^2$

⓮ $-9x^2 + 6x^2 = -x^2$

⓯ $8x^2 - 9x = -4x^2 + 4x$

⓰ $-3x^2 - x = -4x^2 + 5x^2$

⓱ $-x - 7x^2 = 8x^2 + 7x$

⓲ $-3x^2 + 5x^2 = -4 + 7x^2$

❶ $5x + 2x^2 = 4x^2 - 9x$

❷ $2x^2 + 3x = 4x^2 - 7x^2$

❸ $7x^2 + 3x = -4x^2 - x^2$

❹ $5x^2 + x = 8x^2 + 8x^2$

❺ $3x^2 + 4x^2 = -3x - 4x^2$

❻ $9x = -8x^2$

❼ $x^2 - 4x = -6x^2 + 3x$

❽ $9x^2 = 6$

❾ $7x^2 + 8x = -8x + 4x^2$

❿ $-8x^2 + 6x^2 = -x^2 - 7x$

⓫ $5x^2 + 9 = 8x^2 + 9$

⓬ $-7x^2 + 2x = -4x^2 + 2x^2$

⓭ $4x^2 + 7x = 5x^2$

⓮ $-8x^2 + 2x = -5x^2$

⓯ $7x = 6x^2 + 7x$

⓰ $x = -7x^2 + 4x^2$

⓱ $6x^2 - 4x = 4x^2 - 7x$

⓲ $-5 + 8x^2 = -5 + 6x^2$

❶ $8x^2 + 5x^2 = -5x^2 - x$

❷ $8x^2 + 4 = 5x^2 + 5$

❸ $6x^2 - 8x = 8x + x^2$

❹ $x^2 - 6 = 6 - 8x^2$

❺ $x^2 + 7 = 6x^2 + 3$

❻ $-8x - 6x^2 = 5x^2 + 9x$

❼ $x^2 - 8x = -8x - 9x^2$

❽ $-6x - 9x^2 = 2x^2 + 3x$

❾ $-9 + 6x^2 = 5x^2 + 1$

❿ $-2x^2 + 6x = 8x^2 - 6x^2$

⓫ $-6x^2 = x^2$

⓬ $5x - 3x^2 = -7x^2 - 2x$

⓭ $-3x^2 + x = -2x - 6x^2$

⓮ $2x^2 + 6x = -8x + 6x^2$

⓯ $-x^2 + 9x = -6x^2 - 8x^2$

⓰ $7 - 9x^2 = -5 + 4x^2$

⓱ $-8x^2 - 9x = x^2 + 7x$

⓲ $-3x^2 + 6x = 7x^2 - 9x^2$

❶ $-4 + 9x^2 = 1 + 7x^2$

❷ $-7x^2 - 8x = 8x^2 - 4x$

❸ $x^2 = -6x^2 - 5x$

❹ $8x^2 + 7x^2 = 8 + 2x^2$

❺ $-9x^2 - 8x = 5x^2 + 5x$

❻ $-5x + x^2 = -x^2 + 7x$

❼ $3x - 6x^2 = 6x + 6x^2$

❽ $-3x^2 = -7$

❾ $6x^2 + 3x^2 = 3x^2 - 8x$

❿ $-5x^2 + 5x = -9x^2 - 5x$

⓫ $7 = 9x^2 - 3x^2$

⓬ $5x + 3x^2 = 9x^2 + x$

⓭ $-6x + 4x^2 = -6x^2$

⓮ $7x^2 + 3x^2 = 1 + 4x^2$

⓯ $-3x^2 - 4 = -3 - 4x^2$

⓰ $-6x^2 + 7x = -7x^2 + 6x^2$

⓱ $-4x = 8x^2 + x^2$

⓲ $-5x^2 + x^2 = x + 4x^2$

❶ $2x + 7x^2 = -7x^2$

❷ $-4x + 7x^2 = 8x^2 - 7x$

❸ $x - 2x^2 = 5x - 4x^2$

❹ $-1 + 7x^2 = 7 + 2x^2$

❺ $-1 - 9x^2 = -4x^2 - 8$

❻ $9x + 9x^2 = -3x^2 - 8x$

❼ $-6x^2 = 9x^2 + 5x$

❽ $-4x^2 - x = 6x^2 + 4x$

❾ $-2x^2 - 3 = -5$

❿ $-4x^2 - 8x^2 = -6x - 8x^2$

⓫ $-3x = -4x + 2x^2$

⓬ $8x^2 + 5x^2 = 5 + 7x^2$

⓭ $-3 + 8x^2 = 2 - 8x^2$

⓮ $8x^2 - 1 = -9x^2 - 8x^2$

⓯ $-6x^2 + 6 = 6x^2 - 1$

⓰ $2x^2 - 6x^2 = x^2$

⓱ $1 - 2x^2 = -8 + 9x^2$

⓲ $8x^2 - 9 = x^2 - 6x^2$

❶ $5x^2 + x^2 = -2x^2 + 4x$

❷ $-x = x + x^2$

❸ $5x + 2x^2 = x$

❹ $2x + 4x^2 = -4x^2 - 6x$

❺ $-2x^2 - 8x^2 = 2x - x^2$

❻ $-8x + 4x^2 = 2x^2 - 4x^2$

❼ $8 - 6x^2 = 7x^2 - 8x^2$

❽ $6x^2 + 2x^2 = 6x^2 + 2$

❾ $-5x^2 + 5x^2 = 5x^2 - 3x$

❿ $x + 8x^2 = 7x^2 + 9x$

⓫ $-x^2 - 6x^2 = -8x^2 + 1$

⓬ $x^2 - 3x = -3x^2 - 9x$

⓭ $8 = 1 + 5x^2$

⓮ $5x - x^2 = 7x^2 + 8x$

⓯ $2x^2 - 1 = -6x^2 + 5x^2$

⓰ $-3x^2 + 5x = -4x^2 + 5x^2$

⓱ $-8x - 3x^2 = -2x^2 - 5x$

⓲ $-2x^2 + 9x = 2x + 4x^2$

❶ $x^2 - 9x = x^2$

❷ $-6 + 6x^2 = -4x^2 + 6x^2$

❸ $3x^2 = x^2 + 8$

❹ $8x^2 + 8 = x^2 + 9$

❺ $-8 = -2x^2 + 6$

❻ $9x^2 = -x$

❼ $2x + 8x^2 = 3x + 6x^2$

❽ $4x^2 - 5x^2 = 3x^2 - 5$

❾ $-5 + 2x^2 = 4 - 9x^2$

❿ $-7x^2 + 5 = -6x^2 + x^2$

⓫ $-3 - 8x^2 = -7x^2 - 3$

⓬ $-6x - 4x^2 = 2x$

⓭ $8x^2 + 2x = 5x^2$

⓮ $4x^2 + 3x = 4x^2 - 6x^2$

⓯ $-7x - 4x^2 = -8x$

⓰ $5x^2 + 5x^2 = 8x^2 + 1$

⓱ $-4x^2 + 8x = -8x + 2x^2$

⓲ $-9x^2 + 6x^2 = 9x - 5x^2$

Chapter 4: Factoring Quadratic Equations

Quadratic equations – that is, equations that can be expressed in the form $ax^2 + bx + c = 0$ – can be factored. In the previous chapter, we looked at simple cases where either b or c was zero (that is, either linear or constant terms were lacking). In the next chapter, we will learn how to solve the general quadratic equation. In the present chapter, we will look at rational solutions to the quadratic equation. In this case, the quadratic equation can be factored in the form $(dx + e)(fx + g) = 0$.

Before we discuss how to factor the quadratic equation, we review the basics of the foil method. Foil is an abbreviation: First, Outer, Inner, Last. When you multiply an expression of the form $(dx + e)(fx + g)$, first you distribute the dx and the e: $dx(fx + g) + e(fx + g)$. Then you distribute again: $dfx^2 + dgx + efx + eg$. The term dfx^2 arises by multiplying first terms of the factors $(dx + e)$ and $(fx + g)$. The term dgx comes from multiplying the outer terms. The term efx is from the inner terms. The term eg results from the last terms. This way, if you remember that foil stands for First, Outer, Inner, Last, it will help you multiply an expression of the form $(dx + e)(fx + g)$.

When we multiply an expression of the form $(dx + e)(fx + g)$, we obtain a quadratic equation of the form $ax^2 + bx + c = 0$. In this case, we identify a with df, b with dg + ef, and c with eg (simply by comparing $dfx^2 + dgx + efx + eg$ to $ax^2 + bx + c$, realizing that the coefficients of the powers must correspond for the two expressions to be equal). A quadratic equation that has a rational solution can thus be factored into the form $(dx + e)(fx + g) = 0$. As in the last chapter, we see that factoring is essentially the reverse of distributing.

Consider the quadratic equation $3x^2 + 7x + 4 = 0$. It can be factored as $(3x + 4)(x + 1) = 0$. You can (should!) verify this by multiplying $(3x + 4)(x + 1)$ out using the foil method. The main question of this chapter is how to determine the factors – in this case $(3x + 4)(x + 1)$ – given the quadratic equation – in this case, $3x^2 + 7x + 4 = 0$. The answer begins with the numerical factors of the coefficient of the quadratic term and the constant term – in this case, the 3 and the 4. The 3 can be factored as (3)(1), while the 4 can be factored as (2)(2) or (4)(1). This gives the following options: $(3x + 1)(x + 4)$, $(3x + 4)(x + 1)$, and $(3x + 2)(x + 2)$. In addition, either may be factored with two minus signs, as in $(-3)(-1)$, which allows for options like $(-3x + 1)(-x + 4)$ or $(3x - 1)(x - 4)$. Of these possibilities, only one combination of factors and signs will yield the correct signs and values for all three terms. In this case, only $(3x + 4)(x + 1)$ equals $3x^2 + 7x + 4$ when you multiply it out using the foil method.

It is helpful to examine the signs. For example, if the coefficient is positive, then its factors must both be positive or both be negative. For example, in $5x^2$, the 5 can be factored as (5)(1) or $(-5)(-1)$. If instead the coefficient is negative, one of its factors must be positive and the other negative. For example, in $-3x^2$, the -3 can be factored as $(3)(-1)$ or $(-3)(1)$. The linear term comes from the inner and outer terms of the foil method. Factors of the form $(2x + 4)(x + 8)$ yield a positive cross term, factors such as $(3x - 2)(2x - 3)$ yield a negative cross term, and if the signs are mixed you need to use the foil method to see whether the cross term is positive or negative. For example, $(3x - 2)(2x + 4)$ has a cross term of $8x$, while $(2x - 8)(x + 3)$ has a cross term of $-2x$. First think of the factors of the quadratic coefficient and the constant term, then consider the possible signs, and then look at the value and sign of the coefficient of the linear term to determine what combination results in the correct cross term.

The following strategy applies to the quadratic equations of this chapter, where the solutions are rational. Not all quadratic equations have rational solutions, and so quadratic equations cannot always be factored this way. We will learn how to solve the general quadratic equation in the following chapter. When the solution is rational, however, the technique of this chapter is generally most efficient (especially for students who are proficient in the useful technique of factoring). The exercises of this chapter provide good practice in the art of factoring algebraic expressions.

- Move all terms to the same side of the equation using the same technique that we have previously applied: Subtract a positive term from both sides, but add a negative term to both sides. For example, $6x^2 - 21x = -9$ becomes $6x^2 - 21x + 9 = 0$. One side of the equation will equal zero.
- Make a list of the possible numerical factors of the coefficient of the quadratic term and the constant term. For example, in the equation $6x^2 - 21x + 9 = 0$, the 6 can be factored as (1)(6), (6)(1), (2)(3), or (3)(2) and the 9 can be factored as (1)(9), (9)(1), or (3)(3).
- We need to match one pair from the first list with one pair from the second list to make the factors. For example, pairing (1)(6) with (1)(9) yields $(\pm x \pm 1)(\pm 6x \pm 9)$. The $\pm$ means that we are just concentrating on the numbers presently; we will explore the signs later. If you multiply this out with the foil method, you will see that the possibilities are $\pm 6x^2 \pm 15x \pm 9$ and $\pm 6x^2 \pm 3x \pm 9$. These give the correct first and last coefficients, but not the correct cross term (which needs to be $-21x$).
- List all of the possible structures that you can make by matching one pair from each list. For the lists above, the possible structures are $(\pm x \pm 1)(\pm 6x \pm 9)$, $(\pm x \pm 9)(\pm 6x \pm 1)$, $(\pm x \pm 3)(\pm 6x \pm 3)$, $(\pm 2x \pm 1)(\pm 3x \pm 9)$, $(\pm 2x \pm 9)(\pm 3x \pm 1)$, and $(\pm 2x \pm 3)(\pm 3x \pm 3)$.
- Multiply these possible structures out until you find the one that produces the correct quadratic equation – in this case, we are trying to make $6x^2 - 21x + 9 = 0$. Once you obtain a correct answer, stop and go onto the next step. Using the foil method, the inner and outer terms for the possible structures are, for this example: $\pm 6x \pm 9x$, $\pm 54x \pm 1x$, $\pm 18x \pm 3x$, $\pm 3x \pm 18x$, $\pm 27x \pm 2x$, and $\pm 9x \pm 6x$.
- Inner and outer terms of $\pm 18x \pm 3x$ can make the desired cross term of $-21x$, which corresponds to the structure $(\pm x \pm 3)(\pm 6x \pm 3)$. Now play with the signs a little to see which combination of signs works. In this case, it is $(x - 3)(6x - 3) = 0$.
- Set the product of the factors equal to zero, as in the form $(x - 3)(6x - 3) = 0$. Set each factor equal to zero and solve for x using the method of Chapter 1. Here, $x - 3 = 0$ leads to $x = 3$ and $6x - 3 = 0$ leads to $x = 1/2$. The answers are $x = 1/2$ or $x = 3$.

Example 1: $2x^2 - 4 = -2x$. First, add 2x to both sides of the equation so that all of the terms will be on the same side: $2x^2 + 2x - 4 = 0$. The 2 can only factor as (2)(1) or (1)(2), while the 4 can factor as (1)(4), (4)(1), or (2)(2). The possible structures are $(\pm 2x \pm 1)(\pm x \pm 4)$, $(\pm 2x \pm 4)(\pm x \pm 1)$, and $(\pm 2x \pm 2)(\pm x \pm 2)$. Note, for example, that $(\pm x \pm 1)(\pm 2x \pm 4)$ would be the same as $(\pm 2x \pm 4)(\pm x \pm 1)$, which is already on the list. Note also that each of these possible structures produces the correct first and last terms – that is, $2x^2$ and -4. Use the foil method to see what inner and outer terms are possible from these structures: $\pm x \pm 8x$, $\pm 4x \pm 2x$, and $\pm 2x \pm 4x$. The combination $\pm 4x \pm 2x$ can make a cross term of 2x (so could $\pm 2x \pm 4x$). Try different possible signs until you find the combination that works: $(2x + 4)(x - 1) = 0$. Set $2x + 4 = 0$ and $x - 1 = 0$. For $2x + 4 = 0$, subtract 4 from both sides to get $2x = -4$ and then divide both sides by 2 to find $x = -2$. For $x - 1 = 0$, add 1 to both sides to get $x = 1$. The solutions are $x = -2$ and $x = 1$.

Example 2: $3x^2 = 2x + 5$.

$3x^2 - 2x - 5 = 0$

3 = (1)(3) or (3)(1)

5 = (1)(5) or (5,1)

$(\pm x \pm 1)(\pm 3x \pm 5)$, $(\pm x \pm 5)(\pm 3x \pm 1)$

$\pm 3x \pm 5x$ or $\pm 15x \pm x$

$(x + 1)(3x - 5) = 0$

$x + 1 = 0$, $3x - 5 = 0$

$x = -1$, $3x = 5$

$x = -1$, $x = 5/3$

$x = -1, 5/3$

Check: $3x^2 = 3(-1)^2 = 3$.

$2x + 5 = 2(-1) + 5 = 3$. ✓

$3x^2 = 3(5/3)^2 = 25/3$.

$2x + 5 = 2(5/3) + 5 = 10/3 + 5 = 25/3$. ✓

Example 3: $4x^2 + 18x = 10$.

$4x^2 + 18x - 10 = 0$

4 = (1)(4), (4)(1), or (2)(2)

10 = (1)(10), (10,1), (2)(5), or (5)(2)

$(\pm x \pm 1)(\pm 4x \pm 10)$, $(\pm x \pm 10)(\pm 4x \pm 1)$, $(\pm x \pm 2)(\pm 4x \pm 5)$, $(\pm x \pm 5)(\pm 4x \pm 2)$, $(\pm 2x \pm 1)(\pm 2x \pm 10)$, or $(\pm 2x \pm 2)(\pm 2x \pm 5)$

$\pm 4x \pm 10x$, $\pm 40x \pm x$, $\pm 8x \pm 5x$, $\pm 20x \pm 2x$, $\pm 2x \pm 20x$, or $\pm 4x \pm 10x$

$(x + 5)(4x - 2) = 0$

$x + 5 = 0$, $4x - 2 = 0$

$x = -5$, $4x = 2$

$x = -5$, $x = 1/2$

$x = -5, 1/2$

Check: $4x^2 + 18x = 4(-5)^2 + 18(-5)$.

$= 100 - 90 = 10$. ✓

$4x^2 + 18x = 4(1/2)^2 + 18(1/2)$.

$= 1 + 9 = 10$. ✓

Here are a few more notes:

- Why do all of the terms need to be on the same side? Suppose you had $(x + 1)(x - 2) = 3$. You might try $x + 1 = 1$ and $x - 2 = 3$, thinking that (1)(3) = 3, but you can also have $x + 1 = 6$ and $x - 2 = 1/2$, and an infinite number of other combinations. If instead one side equals zero, you know that one factor is zero, removing the ambiguity.
- In principle, you could factor with fractions, like $(3x/4 + 1/2)(x/5 - 2/3)$. However, the exercises in this chapter will factor with integers (but the answers may be fractions).
- The cross term can vanish. For example, $(3x + 2)(3x - 2) = 9x^2 - 4$.
- It is possible to get a double root. For example, $(x - 2)(x - 2) = 0$ has only one answer.

❶ $8x^2 = 12x - 4$

❷ $0 = 15x^2 + 39x - 18$

❸ $-24 = -x^2 + 2x$

❹ $0 = -5x^2 + 18x + 35$

❺ $0 = 8x^2 + 26x + 15$

❻ $-4x^2 - 48 = -32x$

❼ $-2x^2 - 5x - 2 = 0$

❽ $-6x^2 = 41x + 63$

❾ $12x^2 = 3x + 15$

❿ $-4x + 12 = 5x^2$

⓫ $4x^2 - 42x + 54 = 0$

⓬ $4x^2 - 21 = -5x$

❶ $-x^2 + 6x = -16$

❷ $-5x^2 + 21x = 18$

❸ $20x^2 = -42x - 16$

❹ $-16x^2 + 4x = -42$

❺ $-9x^2 = -24x + 15$

❻ $-35 = 10x^2 + 45x$

❼ $9x^2 + 3x - 12 = 0$

❽ $-27 = -15x^2 - 36x$

❾ $-27x = -12x^2 + 27$

❿ $-5x^2 + 6 = -29x$

⓫ $8x^2 + 38x + 35 = 0$

⓬ $15x^2 - 22x = 48$

❶ $-8 = -3x^2 + 23x$

❷ $-22x - 8 = 5x^2$

❸ $-19x - 42 = 2x^2$

❹ $-3x - 56 = -20x^2$

❺ $-56 = -6x^2 - 10x$

❻ $-13x = -20x^2 + 72$

❼ $x - 2 = -6x^2$

❽ $26x + 56 = 10x^2$

❾ $-20x^2 = -57x + 27$

❿ $0 = 3x^2 + 8x + 4$

⓫ $-4x^2 = -18x + 14$

⓬ $0 = -4x^2 + 25x - 36$

❶ $-2x - 16 = -3x^2$

❷ $-4 = -4x^2 - 6x$

❸ $20x^2 + 57x + 40 = 0$

❹ $-20x^2 = 18x - 14$

❺ $12x^2 - 16 = 4x$

❻ $15x^2 + 2 = 11x$

❼ $15x + 63 = 12x^2$

❽ $6x^2 + 18 = 21x$

❾ $1 = x^2$

❿ $3 = 2x^2 - x$

⓫ $-6x^2 - 18x = 12$

⓬ $4x^2 + 25x = -6$

❶ $-40\,x = 8\,x^2 + 42$

❷ $-x^2 - 14\,x - 49 = 0$

❸ $20\,x = 25\,x^2 - 21$

❹ $-8\,x^2 - 30 = 34\,x$

❺ $4\,x^2 - 4 = 0$

❻ $39\,x = 15\,x^2 + 24$

❼ $-4\,x^2 - 29\,x - 30 = 0$

❽ $-15\,x^2 - 40 = 49\,x$

❾ $-42\,x - 9 = -15\,x^2$

❿ $-6\,x + 9 = 3\,x^2$

⓫ $-32 = 3\,x^2 - 20\,x$

⓬ $-3\,x^2 = 26\,x + 16$

❶ $0 = -20x^2 - 46x - 24$

❷ $20x^2 - 11x = 42$

❸ $0 = -12x^2 - 27x - 15$

❹ $10x^2 + 42 = -44x$

❺ $-6 = 16x^2 + 28x$

❻ $20x^2 - 4 = -16x$

❼ $29x = 5x^2 + 20$

❽ $8x - 8 = 2x^2$

❾ $-x^2 = 4x - 5$

❿ $-9x - 54 = -9x^2$

⓫ $-15x^2 - 6 = -21x$

⓬ $32 = 10x^2 + 4x$

❶ $0 = 5x^2 - 25x - 30$

❷ $x = -10x^2 + 24$

❸ $0 = -15x^2 + 3x + 54$

❹ $-20x^2 = -20$

❺ $-5x^2 + 26x = -24$

❻ $-10x^2 - 35x = -45$

❼ $6x^2 = 3x + 30$

❽ $5x^2 - 25x - 30 = 0$

❾ $5x^2 - 7 = -34x$

❿ $9x^2 = 64$

⓫ $3x^2 = 9x - 6$

⓬ $-2x^2 = 14x + 12$

❶ $6x^2 - 3 = 7x$

❷ $19x = 6x^2 - 36$

❸ $-20x - 16 = -6x^2$

❹ $5x^2 - 10x = -5$

❺ $-8x^2 + 6 = -8x$

❻ $10x^2 = 43x - 45$

❼ $12x^2 + 30x + 12 = 0$

❽ $-3x^2 + 18x = 24$

❾ $10 = 12x^2 - 14x$

❿ $3x^2 + 10x = -7$

⓫ $-x^2 - 4 = -5x$

⓬ $12x^2 + 34x + 14 = 0$

❶ $-4\,x = -12\,x^2 + 8$

❷ $-5\,x^2 - 5 = 26\,x$

❸ $-15\,x^2 = -42\,x + 24$

❹ $-12\,x^2 - 33\,x - 21 = 0$

❺ $0 = 9\,x^2 + 21\,x - 8$

❻ $4\,x^2 + 15\,x = 25$

❼ $17\,x = -4\,x^2 - 15$

❽ $9\,x^2 + 30\,x + 9 = 0$

❾ $0 = 8\,x^2 + 14\,x + 3$

❿ $-25\,x^2 + 1 = 0$

⓫ $-12\,x^2 - 10\,x = -8$

⓬ $15\,x^2 = -9\,x + 24$

❶ $4x^2 + 15 = 23x$

❷ $-13x + 21 = -2x^2$

❸ $-52x = -10x^2 - 48$

❹ $-64 = 20x^2 - 72x$

❺ $-20x^2 - 46x - 18 = 0$

❻ $33x = -4x^2 - 8$

❼ $10x^2 = x + 24$

❽ $12x^2 - 13x - 35 = 0$

❾ $-21x + 45 = 12x^2$

❿ $17x - 12 = -5x^2$

⓫ $-4x^2 + 9 = -16x$

⓬ $0 = 4x^2 - 10x - 6$

❶ $-2x^2 - 4x + 48 = 0$

❷ $5x^2 + 9 = 18x$

❸ $-10x^2 + 5x + 5 = 0$

❹ $-20x^2 = -18x + 4$

❺ $4x^2 + 16x = 48$

❻ $-68x + 72 = -16x^2$

❼ $-3x = 2x^2 - 14$

❽ $16x^2 - 32 = -16x$

❾ $-12x = -2x^2 - 18$

❿ $4x^2 - 36 = 10x$

⓫ $12x = 9x^2 - 21$

⓬ $54x = -20x^2 - 36$

❶ $-12x^2 + 12x + 24 = 0$

❷ $0 = -3x^2 + 23x - 14$

❸ $-25x^2 + 7 = -30x$

❹ $46x + 9 = -5x^2$

❺ $16x + 4 = 20x^2$

❻ $0 = -15x^2 - 20x - 5$

❼ $12x^2 + 7 = 25x$

❽ $9x^2 = -48x - 63$

❾ $3x + 35 = 20x^2$

❿ $-13x = 4x^2 + 9$

⓫ $61x - 72 = 10x^2$

⓬ $2x^2 - 13x = -21$

❶ $8x^2 = 10x - 2$

❷ $31x = 20x^2 - 9$

❸ $-63 = -x^2 + 2x$

❹ $-40 = -10x^2 - 30x$

❺ $-4x^2 + 6x + 4 = 0$

❻ $-12x + 16 = -2x^2$

❼ $-26x - 24 = -5x^2$

❽ $0 = 5x^2 + 14x + 9$

❾ $-2x^2 + 9x + 81 = 0$

❿ $-64x = 15x^2 + 64$

⓫ $5x^2 - 12 = -17x$

⓬ $0 = 3x^2 + 22x - 16$

❶ $12x^2 - 26x = 16$

❷ $34x + 9 = 8x^2$

❸ $0 = -2x^2 - x + 36$

❹ $-6x^2 + 16 = 4x$

❺ $12x^2 = -36x - 24$

❻ $11x + 6 = -3x^2$

❼ $12x^2 = 48$

❽ $-9x^2 = 27x + 14$

❾ $-2x^2 - 5x = 2$

❿ $32 = -6x^2 - 28x$

⓫ $6x^2 + 23x = -20$

⓬ $-2x^2 = -10x - 48$

❶ $5x^2 - 24x + 16 = 0$

❷ $28 = -3x^2 - 25x$

❸ $-x - 72 = -x^2$

❹ $25x^2 = 55x - 18$

❺ $15x^2 - 15 = 16x$

❻ $-7x = -5x^2 - 2$

❼ $18 = 12x^2 + 19x$

❽ $56 = 20x^2 - 12x$

❾ $15x^2 - 47x = -36$

❿ $63 = -10x^2 + 53x$

⓫ $-14 = 6x^2 - 20x$

⓬ $2x^2 - 25x = -63$

❶ $6x^2 = 15x + 21$

❷ $-34x + 24 = -12x^2$

❸ $-4x^2 + 32x - 48 = 0$

❹ $36x = -5x^2 + 32$

❺ $-40 = 15x^2 + 55x$

❻ $3x^2 + 36x + 81 = 0$

❼ $16x = 4x^2 + 15$

❽ $-34x - 40 = 6x^2$

❾ $7x - 49 = -20x^2$

❿ $8x^2 = -14x - 5$

⓫ $6x - 16 = -x^2$

⓬ $-18 = -15x^2 + 21x$

❶ $2x^2 - x - 1 = 0$

❷ $x^2 = -17x - 72$

❸ $21x = 3x^2 + 30$

❹ $-8x^2 + 12x = -20$

❺ $12x^2 - 30 = 2x$

❻ $25x^2 - 80x = -64$

❼ $-19x = -12x^2 + 21$

❽ $-8x^2 = 14x - 72$

❾ $-32x - 12 = 16x^2$

❿ $8x = -x^2 - 12$

⓫ $6x^2 = -19x - 15$

⓬ $-13x - 35 = -4x^2$

❶ $-20x^2 + 39x = 7$

❷ $8x = 12x^2 - 64$

❸ $-2x^2 - 12x = -54$

❹ $-8x = 3x^2 - 16$

❺ $2x^2 + 8x - 42 = 0$

❻ $-24 = 12x^2 + 34x$

❼ $-35 = -16x^2 - 8x$

❽ $17x = -3x^2 - 24$

❾ $-20x^2 + x = -63$

❿ $-26x - 12 = 12x^2$

⓫ $-12x^2 + 21x = 9$

⓬ $-3x^2 + 21 = -18x$

❶ $5x^2 = 21x - 18$

❷ $-37x + 18 = -15x^2$

❸ $6x = 8x^2 + 1$

❹ $-6x^2 = -3x - 63$

❺ $x = -10x^2 + 3$

❻ $5x^2 + 8 = 41x$

❼ $5 = 6x^2 + 13x$

❽ $37x - 14 = 5x^2$

❾ $-23x = -3x^2 + 36$

❿ $10x^2 + 7x - 12 = 0$

⓫ $-20x^2 + 12 = -x$

⓬ $0 = -3x^2 + 18x + 48$

❶ $-18 = 8x^2 + 40x$

❷ $4x^2 = 20x - 16$

❸ $-3x^2 = 12x + 12$

❹ $0 = -16x^2 - 8x + 15$

❺ $0 = 8x^2 - 14x + 3$

❻ $49 = -3x^2 - 28x$

❼ $-20x^2 - 18 = 53x$

❽ $37x = 12x^2 + 28$

❾ $-35x - 25 = 10x^2$

❿ $-6x^2 + 19x + 7 = 0$

⓫ $6x^2 + x = 15$

⓬ $-17x = 4x^2 - 15$

❶ $0 = -20x^2 - 6x + 54$

❷ $x^2 + 25 = -10x$

❸ $-5x^2 + x + 18 = 0$

❹ $-4x^2 - 72 = 41x$

❺ $-4x = 3x^2 - 32$

❻ $-15x^2 - 9 = 48x$

❼ $0 = -10x^2 + 27x + 28$

❽ $48 = 5x^2 + 22x$

❾ $-8x^2 - 5 = 22x$

❿ $-20 = 4x^2 + 21x$

⓫ $0 = 8x^2 - 34x - 9$

⓬ $-20x^2 + 3x = -2$

❶ $-5x^2 + 38x + 16 = 0$

❷ $-4x^2 - 4 = -8x$

❸ $0 = -5x^2 - 7x + 24$

❹ $0 = -3x^2 + 11x - 10$

❺ $8x^2 - 36 = -2x$

❻ $5x^2 + 53x + 72 = 0$

❼ $-4x^2 + 37x - 9 = 0$

❽ $0 = 3x^2 - 17x - 28$

❾ $-2x^2 - 45 = 19x$

❿ $-8x^2 - 30x = 27$

⓫ $25x^2 + 10x - 24 = 0$

⓬ $-14 = 9x^2 - 27x$

❶ $0 = -20x^2 + 37x - 15$

❷ $8x - 1 = 15x^2$

❸ $-14 = 15x^2 + 41x$

❹ $6x^2 + 28x = -30$

❺ $-10x^2 + 18 = 8x$

❻ $48 = 5x^2 - 34x$

❼ $10x + 18 = 8x^2$

❽ $0 = 20x^2 - 17x - 24$

❾ $6x^2 - 14x + 4 = 0$

❿ $-16x^2 + 27 = 24x$

⓫ $0 = -6x^2 + 30x - 36$

⓬ $-25x^2 = -40x + 12$

❶ $3 = 16x^2 - 8x$

❷ $-18x = 20x^2 - 18$

❸ $-9 = 10x^2 - 47x$

❹ $-6x^2 - x + 1 = 0$

❺ $3x^2 - 28 = -17x$

❻ $-12x = -x^2 - 32$

❼ $3x^2 = -10x - 7$

❽ $8x^2 - 18x + 9 = 0$

❾ $25x^2 = 5x + 2$

❿ $10x^2 + 7 = 19x$

⓫ $8x^2 + 2x - 21 = 0$

⓬ $-3x - 18 = -10x^2$

❶ $-2x^2 = 5x + 2$

❷ $-12x^2 = 34x + 24$

❸ $0 = 12x^2 + 40x + 32$

❹ $-48x = -20x^2 - 28$

❺ $0 = 25x^2 + 10x - 15$

❻ $-6x^2 = -39x + 54$

❼ $-x = 6x^2 - 35$

❽ $5x^2 + 43x = -24$

❾ $29x = -5x^2 + 6$

❿ $23x = 5x^2 + 12$

⓫ $4x^2 - 6x = 10$

⓬ $2x + 28 = 8x^2$

❶ $0 = 8x^2 - 40x + 32$

❷ $-6 = -25x^2 + 25x$

❸ $-10x^2 - 27 = 33x$

❹ $0 = 10x^2 - 24x + 8$

❺ $4x^2 - 7x - 36 = 0$

❻ $-27x = 5x^2 + 10$

❼ $12x^2 - 59x + 72 = 0$

❽ $4x + 7 = 3x^2$

❾ $18x = -8x^2 - 7$

❿ $-25x = 4x^2 + 6$

⓫ $-20x^2 + 13x + 15 = 0$

⓬ $-13x = 3x^2 - 10$

❶ $-16x^2 + 52x - 40 = 0$

❷ $-27x + 81 = 10x^2$

❸ $0 = 8x^2 + 18x + 7$

❹ $15x^2 = 18x - 3$

❺ $-20x^2 = 14x + 2$

❻ $0 = 9x^2 - 36$

❼ $-14x - 16 = 3x^2$

❽ $12x^2 - 16 = 16x$

❾ $-16x^2 = 36x + 14$

❿ $-9x = -3x^2 - 6$

⓫ $-3x^2 + 45 = 6x$

⓬ $15x^2 - 18 = 17x$

❶

$-4 = -4x^2 - 15x$

❷

$0 = 6x^2 - 17x - 45$

❸

$16 = 8x^2 - 28x$

❹

$-64x - 48 = 20x^2$

❺

$-3x^2 = -8x - 16$

❻

$54 = 8x^2 + 6x$

❼

$5x^2 + 35x = -30$

❽

$10x = 12x^2 - 42$

❾

$-5x^2 + 41x = 8$

❿

$-26x + 18 = -8x^2$

⓫

$-2x^2 - 9x + 18 = 0$

⓬

$-20x^2 + 36 = -16x$

❶ $-48\,x = -20\,x^2 - 16$

❷ $5\,x^2 + 36 = 36\,x$

❸ $20 = -12\,x^2 - 32\,x$

❹ $-15\,x^2 + 19\,x = 6$

❺ $0 = -10\,x^2 + 18\,x - 8$

❻ $-5\,x^2 - 18 = -21\,x$

❼ $0 = 20\,x^2 - 47\,x + 24$

❽ $-38\,x + 24 = -10\,x^2$

❾ $-12\,x^2 + 23\,x - 10 = 0$

❿ $16\,x - 3 = 5\,x^2$

⓫ $0 = 25\,x^2 - 1$

⓬ $6\,x^2 + 42 = -32\,x$

❶

$0 = -4x^2 + 11x + 45$

❷

$3x^2 + 24 = 18x$

❸

$4x^2 - 28x - 32 = 0$

❹

$64 = 4x^2$

❺

$-20x^2 + 35x - 15 = 0$

❻

$8x^2 + 6x - 2 = 0$

❼

$-27 = -15x^2 + 36x$

❽

$-15x^2 - 11x = -56$

❾

$0 = -20x^2 - 8x + 12$

❿

$5x^2 + 36x - 32 = 0$

⓫

$25x^2 = -25x - 6$

⓬

$-5x^2 - 13x + 6 = 0$

Chapter 5: Formula for the Quadratic Equation

In this chapter, we will learn that there is a formula – called the quadratic formula – for solving the quadratic equation – that is, an equation that can be cast in the form $ax^2 + bx + c = 0$, where a, b, and c are constants.

We will first derive the quadratic formula, and then discuss how to apply it. You can still solve all of the exercises even if you feel lost trying to follow the derivation below. However, it is worth trying to understand the derivation for a few reasons. For one, it shows where the quadratic equation comes from. For another, mathematical derivations are often very symbolic and very abstract, and the more that you attempt to understand them, the more it benefits your ability to think abstractly and in symbolic terms – even if you only get a small piece of it at first. Nonetheless, if you find it too much, you can always skip down to where the derivation ends and the text describes how to apply the result.

In general, the value of x that solves the quadratic formula may be rational, irrational, or a combination of both – like $x = 2 + \sqrt{3}$. In the most general case, where x is a sum (or difference) of rational and irrational parts, the quadratic equation can be factored in the form $(x + d + \sqrt{e})(x + f + \sqrt{g}) = 0$. If we multiply this out, we get $x^2 + dx + fx + \sqrt{e}\,x + \sqrt{g}\,x + df + d\sqrt{g} + \sqrt{e}\,f + \sqrt{e}\sqrt{g}$.

If we divide both sides of the quadratic equation, $ax^2 + bx + c = 0$, by a, we get $x^2 + bx/a + c/a = 0$. This form of the quadratic equation, $x^2 + bx/a + c/a = 0$, matches the coefficient of the quadratic term in the equation $x^2 + dx + fx + \sqrt{e}\,x + \sqrt{g}\,x + df + d\sqrt{g} + \sqrt{e}\,f + \sqrt{e}\sqrt{g}$. Both forms of the quadratic equation would be identical, in general, if $b/a = d + f + \sqrt{e} + \sqrt{g}$ and if $c/a = df + d\sqrt{g} + \sqrt{e}\,f + \sqrt{e}\sqrt{g}$. This way, the coefficients of the quadratic, linear, and constant terms will match.

Let us assume that a, b, c, d, e, f, and g are all rational numbers. Then b/a is rational. Similarly, d + f is rational. However, $\sqrt{e} + \sqrt{g}$ is irrational. There is no way that a rational number plus an irrational number can equal a rational number. Then the only way that b/a can equal $d + f + \sqrt{e} + \sqrt{g}$ is if $\sqrt{e} + \sqrt{g} = 0$, which means that $\sqrt{e} = -\sqrt{g}$. This means that the quadratic equation factors as $(x + d + \sqrt{e})(x + f - \sqrt{e}) = 0$. This matches the quadratic equation, $x^2 + bx/a + c/a = 0$, if $b/a = d + f$ and if $c/a = df - d\sqrt{e} + \sqrt{e}\,f - e$, where the last term came from $\sqrt{e}\sqrt{g} = -\sqrt{e}\sqrt{e} = -e$. Again, the irrational part must cancel in order for c/a to be rational: $-d\sqrt{e} + \sqrt{e}\,f = 0$, which means that $f = d$.

Setting f = d, the factoring becomes $(x + d + \sqrt{e})(x + d - \sqrt{e}) = 0$. This matches the quadratic equation, $x^2 + bx/a + c/a = 0$, if $b/a = d + f = 2d$ and if $c/a = df - e = d^2 - e$. Solving for d, we find that $d = b/(2a)$, then solving for e, we find that $e = d^2 - c/a = [b/(2a)]^2 - c/a$, which simplifies to $e = b^2/(4a^2) - c/a$.

This means that the quadratic equation, in the form $x^2 + bx/a + c/a = 0$, can be factored as $[x + b/(2a) + \sqrt{b^2/(4a)^2 - c/a}\,][x + b/(2a) - \sqrt{b^2/(4a)^2 - c/a}\,] = 0$. Either one factor or the other must be zero: $x + b/(2a) + \sqrt{b^2/(4a)^2 - c/a} = 0$ gives $x = -b/(2a) - \sqrt{b^2/(4a)^2 - c/a}$ as a solution, and $x + b/(2a) - \sqrt{b^2/(4a)^2 - c/a} = 0$ gives $x = -b/(2a) + \sqrt{b^2/(4a)^2 - c/a}$ as a solution. Both solutions can be stated concisely as $x = -b/(2a) \pm \sqrt{b^2/(4a)^2 - c/a}$.

We can express this in standard form by finding a common denominator inside the squareroot: $b^2/(4a)^2 - c/a = b^2/(4a)^2 - 4ac/(4a)^2 = (b^2 - 4ac)/(4a)^2$. The solution to the quadratic equation then becomes $x = -b/(2a) \pm \sqrt{b^2 - 4ac}/(2a)$. Both terms have the same denominator (2a). The solution to the quadratic equation is called the quadratic formula, and is generally expressed in the following form:

$$x = \frac{-b \pm \sqrt{b^2 - 4ac}}{2a}$$

That concludes our derivation of the quadratic formula. You can verify that this solves the quadratic equation, $ax^2 + bx + c = 0$, by plugging it in for x and simplifying. It will be very helpful to memorize the above equation – the quadratic formula. Now we will describe how to apply the quadratic formula to solve the quadratic equation.

If you know the quadratic formula, $x = \frac{-b \pm \sqrt{b^2 - 4ac}}{2a}$, solving the quadratic equation is very straightforward: Mainly, you identify the constants a, b, and c, by comparing your equation with numbers to the symbolic equation $ax^2 + bx + c = 0$, and then you plug these values of a, b, and c into the quadratic formula.

Following is the strategy for how to apply the quadratic formula to solve the general quadratic equation.

- You must first put the quadratic equation in standard form. Move all of the terms to the same side of the equation as usual (that is, if a term is positive, subtract it from both sides in order to move it to the other side, and if the term is negative, instead add it to both sides). Collect like terms together: For example, $3x + 4x$ can be combined into $7x$, and $6x^2 - 4x^2$ can be simplified to $2x^2$. Write the quadratic term first, then the linear term, and lastly the constant term – in that order. You will have three terms added together set equal to zero. The standard form is $ax^2 + bx + c = 0$.
- Identify the coefficients of each term as a, b, and c. Signs are important. For example, in the quadratic equation $3x^2 - 5x - 4 = 0$, $a = 3$, $b = -5$, and $c = -4$.

- Plug these values of a, b, and c into the quadratic formula and, if necessary, simplify. For example, with a = 3, b = –5, and c = –4, the quadratic formula gives: $x = \frac{-b \pm \sqrt{b^2 - 4ac}}{2a} = \frac{-(-5) \pm \sqrt{(-5)^2 - 4(3)(-4)}}{2(3)} = \frac{5 \pm \sqrt{25+48}}{6} = \frac{5 \pm \sqrt{73}}{6}$.

Example 1: $2x^2 - 2 = 3x$. All of the terms will be on the left side of the equation if we move the 3x to the left. Since 3x is positive, we subtract it from both sides of the equation: $2x^2 - 2 - 3x = 0$. We need the quadratic term first, then the linear term, then the constant term, so we reorder the equation as $2x^2 - 3x - 2 = 0$. Comparing this to the quadratic equation $ax^2 + bx + c = 0$, we see that a = 2, b = –3, and c = –2. We plug these values into the quadratic formula: $x = \frac{-b \pm \sqrt{b^2 - 4ac}}{2a} = \frac{-(-3) \pm \sqrt{(-3)^2 - 4(2)(-2)}}{2(2)} = \frac{3 \pm \sqrt{9+16}}{4} = \frac{3 \pm \sqrt{25}}{4} = \frac{3 \pm 5}{4}$. The numerator equals –2 or 8, so x = –2/4 = –1/2 or x = 8/4 = 2. The two answers are x = –1/2 or x = 2. We can check these answers by plugging them back into the original equation: $2x^2 - 2 = 2(-1/2)^2 - 2 = 2/4 - 2 = -3/2$, which agrees with $3x = 3(-1/2) = -3/2$, and $2x^2 - 2 = 2(2)^2 - 2 = 8 - 2 = 6$, which agrees with $3x = 3(2) = 6$.

Example 2: $x^2 + 2x = 3$.

$$x^2 + 2x - 3 = 0$$

$$a = 1, b = 2, c = -3$$

$$x = \frac{-2 \pm \sqrt{2^2 - 4(1)(-3)}}{2(1)} = \frac{-2 \pm \sqrt{4+12}}{2}$$

$$x = \frac{-2 \pm \sqrt{16}}{2} = \frac{-2 \pm 4}{2}$$

$$\boxed{x = -3, 1}$$

Check: $x^2 + 2x = (-3)^2 + 2(-3) = 9 - 6 = 3$. ✓

$x^2 + 2x = (1)^2 + 2(1) = 1 + 2 = 3$. ✓

Example 3: $3x^2 + 2 = 8x$.

$$3x^2 - 8x + 2 = 0$$

$$a = 3, b = -8, c = 2$$

$$x = \frac{-(-8) \pm \sqrt{(-8)^2 - 4(3)(2)}}{2(3)} = \frac{8 \pm \sqrt{64-24}}{6}$$

$$x = \frac{8 \pm \sqrt{40}}{6} = \frac{8 \pm 2\sqrt{10}}{6} = \frac{4 \pm \sqrt{10}}{3}$$

$$\boxed{x = 4/3 \pm \sqrt{10}/3}$$

Check: $3x^2 + 2 = 3(4/3 \pm \sqrt{10}/3)^2 + 2$

$= 16/3 \pm 8\sqrt{10}/3 + 10/3 + 2$

$= 32/3 \pm 8\sqrt{10}/3$.

$8x = 8(4/3 \pm \sqrt{10}/3) = 32/3 \pm 8\sqrt{10}/3$. ✓

- A couple of common, silly mistakes that students tend to make are with the minus signs. Note, for example, that $-(-2) = 2$ and $(-4)^2 = 16$. It's also important to make sure that all of the terms are on the same side of the equation – and in the right order – before you try to identify a, b, and c. When you identify a, b, and c, don't forget to include any minus signs that might be there.
- The inside of the squareroot of the quadratic formula, $b^2 - 4ac$, is called the discriminant. The discriminant tells you a lot about the answer: If $b^2 > 4ac$ there are two real roots, if $b^2 = 4ac$ there is a double root, and if $b^2 < 4ac$ the roots are complex.

❶ $-5x^2 - 2 = 14x$

❷ $3x^2 + 12x - 3 = 0$

❸ $-5 = -3x^2 - 5x$

❹ $-6x - 1 = -x^2$

❺ $2x = 5x^2 - 8$

❻ $-9x - 8 = x^2$

❼ $6 = 5x^2 - 7x$

❽ $1 = 4x^2 + 3x$

❾ $-5 = 5x^2 + 18x$

❿ $-10x - 1 = -4x^2$

⓫ $-x^2 + 15x = 9$

⓬ $3x^2 = -14x + 2$

⓭ $20x = -4x^2 - 5$

⓮ $4x^2 = -11x - 4$

⓯ $5x^2 - 18x = -7$

⓰ $14x = x^2 - 7$

⓱ $0 = 4x^2 + 4x - 3$

⓲ $2x^2 - 9 = -2x$

❶ $-8x = -x^2 + 3$

❷ $2x^2 - 7 = -4x$

❸ $-2 = x^2 + 12x$

❹ $-4x^2 + 1 = -20x$

❺ $-4x^2 + 2 = -11x$

❻ $-6 = -5x^2 - 13x$

❼ $-9 = -2x^2 - 4x$

❽ $-4 = -3x^2 + 7x$

❾ $12x + 8 = x^2$

❿ $16x = 2x^2 - 5$

⓫ $-x^2 + 18x = -3$

⓬ $4x^2 - 8 = -13x$

⓭ $6 = -5x^2 - 20x$

⓮ $-2x^2 - 19x - 9 = 0$

⓯ $x^2 - 16x - 7 = 0$

⓰ $5x^2 = -12x + 1$

⓱ $2x^2 = -14x + 7$

⓲ $0 = 4x^2 + 12x - 7$

❶ $2x^2 = 8x - 7$

❷ $7 = -2x^2 - 16x$

❸ $2x^2 - 12x + 5 = 0$

❹ $0 = x^2 - x - 4$

❺ $8x + 6 = -2x^2$

❻ $-6x - 1 = -x^2$

❼ $-4x = x^2 - 3$

❽ $-x^2 + 5 = x$

❾ $15x + 9 = 5x^2$

❿ $x^2 - 10x - 6 = 0$

⓫ $-6x = -x^2 + 3$

⓬ $9x = -5x^2 + 2$

⓭ $-2x^2 - 9x + 9 = 0$

⓮ $-14x = -5x^2 + 6$

⓯ $-8 = -x^2 - 18x$

⓰ $4x^2 - 14x = 7$

⓱ $-2x^2 - 12x = 3$

⓲ $5 = x^2 - 16x$

❶ $12x - 2 = -4x^2$

❷ $2 = -2x^2 - 16x$

❸ $5x - 1 = 2x^2$

❹ $9 = 5x^2 + 8x$

❺ $-8x + 3 = 4x^2$

❻ $3x^2 - 10x - 7 = 0$

❼ $3 = 5x^2 + 18x$

❽ $0 = -4x^2 + 16x + 7$

❾ $-4x^2 - 9 = -14x$

❿ $3x^2 + 12x - 8 = 0$

⓫ $-3x^2 + 18x - 2 = 0$

⓬ $0 = -3x^2 - 3x + 3$

⓭ $-2x^2 + 16x = -1$

⓮ $-4x^2 - 5 = 13x$

⓯ $3x^2 = 17x + 3$

⓰ $0 = 3x^2 - 8x + 5$

⓱ $4x^2 + 7 = 14x$

⓲ $-4x^2 - 9 = -20x$

❶ $-3x^2 - 14x = 9$

❷ $10x - 1 = 3x^2$

❸ $4x^2 + 6x = 7$

❹ $-x^2 + 5x = -8$

❺ $-2 = -5x^2 - 8x$

❻ $3x^2 - 9 = 2x$

❼ $5x^2 - 18x = -8$

❽ $-3x^2 + 3 = 6x$

❾ $4x^2 - 19x = 5$

❿ $4x^2 - 9 = -4x$

⓫ $x^2 + 6x = -5$

⓬ $-x^2 = 16x - 8$

⓭ $2x^2 + 4x - 8 = 0$

⓮ $-2x^2 = 14x + 6$

⓯ $14x + 7 = 4x^2$

⓰ $-4x^2 = 14x - 8$

⓱ $-2x^2 - 8x = 1$

⓲ $-x^2 - 9x + 1 = 0$

❶ $4x^2 - 10x + 1 = 0$

❷ $14x = 4x^2 + 9$

❸ $2x^2 - 6 = 8x$

❹ $2x^2 - 6x = 4$

❺ $4x^2 - 9 = -18x$

❻ $0 = x^2 - 16x - 8$

❼ $-18x - 5 = -3x^2$

❽ $-2x^2 - 4 = -6x$

❾ $-4x^2 + 9x = 4$

❿ $-3x^2 + 4 = -11x$

⓫ $2x^2 + x - 2 = 0$

⓬ $16x = x^2 - 1$

⓭ $9x = -3x^2 - 1$

⓮ $5x^2 - 16x = -8$

⓯ $-5x^2 + 16x + 5 = 0$

⓰ $2x^2 - 8x = -1$

⓱ $-5x^2 - 15x + 9 = 0$

⓲ $-2 = -4x^2 - 14x$

❶ $-x + 2 = 5x^2$

❷ $-5x^2 + 11x + 7 = 0$

❸ $-5 = 3x^2 - 11x$

❹ $10x + 4 = -x^2$

❺ $-4x^2 + 11x + 3 = 0$

❻ $4x - 4 = -5x^2$

❼ $-2x^2 + 20x = 7$

❽ $2x^2 - 18x + 7 = 0$

❾ $-x^2 - 14x + 1 = 0$

❿ $-x^2 - 3x = -9$

⓫ $-9 = -5x^2 + 18x$

⓬ $6 = 5x^2 - 10x$

⓭ $3 = 3x^2 - 15x$

⓮ $2x^2 + 14x = -9$

⓯ $2x^2 + x - 7 = 0$

⓰ $-x^2 + 8 = -4x$

⓱ $-6 = -4x^2 - 8x$

⓲ $2x^2 - 9x + 1 = 0$

❶ $0 = -x^2 + 3x + 2$

❷ $-10x = 2x^2 + 8$

❸ $16x - 9 = -2x^2$

❹ $2x^2 + 16x = -9$

❺ $-x^2 - 16x = -3$

❻ $4x^2 + 18x + 6 = 0$

❼ $-8 = -x^2 - 5x$

❽ $14x - 1 = 4x^2$

❾ $x^2 = -9x - 3$

❿ $-5x^2 + 4x = -3$

⓫ $1 = -4x^2 - 5x$

⓬ $0 = 5x^2 + 20x + 4$

⓭ $-5x - 7 = -x^2$

⓮ $-2x^2 + 7 = 10x$

⓯ $x^2 + 5x = 4$

⓰ $4x^2 + 10x + 4 = 0$

⓱ $-2x^2 + 14x + 3 = 0$

⓲ $x^2 - 2 = -6x$

❶ $-3x^2 - 4x = -3$

❷ $-17x + 2 = x^2$

❸ $0 = -x^2 - 10x - 9$

❹ $-12x + 8 = 3x^2$

❺ $4 = -3x^2 - 7x$

❻ $2x^2 + 18x + 9 = 0$

❼ $-3x^2 - 1 = -16x$

❽ $5x^2 - 5 = 8x$

❾ $-3x^2 - 2 = 10x$

❿ $0 = 4x^2 + 11x + 4$

⓫ $6x = -3x^2 + 7$

⓬ $5 = -2x^2 + 16x$

⓭ $-4x^2 - 10x - 1 = 0$

⓮ $-5x^2 - 20x - 1 = 0$

⓯ $3x = -2x^2 + 7$

⓰ $16x - 5 = 5x^2$

⓱ $0 = 5x^2 - 16x + 3$

⓲ $x^2 + 12x = 2$

❶ $x^2 + 7x - 1 = 0$

❷ $-4 = x^2 + 7x$

❸ $5x^2 + 9x = 2$

❹ $-x^2 - 7 = -11x$

❺ $-14x + 7 = -4x^2$

❻ $4x^2 - 7 = 14x$

❼ $2 = x^2 - 18x$

❽ $-5x^2 - 10x = -1$

❾ $x^2 + 6x - 1 = 0$

❿ $16x - 8 = -x^2$

⓫ $3x^2 - 3 = -8x$

⓬ $-18x + 1 = -4x^2$

⓭ $-x^2 - 18x - 9 = 0$

⓮ $-x^2 = 13x - 5$

⓯ $-9 = -4x^2 + 6x$

⓰ $-10x = x^2 + 6$

⓱ $-3x^2 + 13x = 9$

⓲ $5x^2 + 16x = -1$

❶ $5 = x^2 - 18x$

❷ $-3x^2 = 16x + 5$

❸ $8 = 2x^2 - 18x$

❹ $3x + 8 = x^2$

❺ $0 = 3x^2 + 12x - 3$

❻ $-5x^2 + 6 = -16x$

❼ $5x = 4x^2 - 2$

❽ $-2x^2 = 8x + 1$

❾ $2x^2 = x + 8$

❿ $5x^2 = 10x - 3$

⓫ $-3x^2 - 18x = -2$

⓬ $3x^2 = 5x + 6$

⓭ $-2x^2 + 2 = 4x$

⓮ $-x^2 = 9x + 9$

⓯ $-2x^2 - 16x - 5 = 0$

⓰ $5x^2 = -14x - 6$

⓱ $x^2 = 12x - 5$

⓲ $x^2 = -7x + 2$

❶ $6 = -4x^2 + 10x$

❷ $-20x = -3x^2 - 6$

❸ $-x^2 - 5x = 5$

❹ $0 = -2x^2 + 16x - 6$

❺ $4x^2 + 14x - 7 = 0$

❻ $13x = 5x^2 + 6$

❼ $0 = -x^2 - 6x + 6$

❽ $-2x = -x^2 + 4$

❾ $-5x^2 + 19x = 5$

❿ $-3x^2 = -x - 5$

⓫ $-13x - 6 = 3x^2$

⓬ $-6x = -5x^2 + 9$

⓭ $-5x^2 + 12x = 1$

⓮ $-2x^2 = 4x + 1$

⓯ $2x^2 + 15x - 9 = 0$

⓰ $-4x^2 + 7x = -2$

⓱ $4x^2 - 10x - 7 = 0$

⓲ $-4x^2 + 2 = -4x$

❶ $0 = -4x^2 + 10x - 3$

❷ $4x^2 - 2 = -5x$

❸ $3x^2 = 7x + 6$

❹ $-4x^2 = -12x + 3$

❺ $8 = 3x^2 + 4x$

❻ $-x^2 - 10x = 4$

❼ $5x^2 - 15x + 9 = 0$

❽ $7x = -x^2 + 9$

❾ $x^2 + 16x = 2$

❿ $-19x + 9 = -x^2$

⓫ $-x^2 = 20x + 2$

⓬ $-10x = -5x^2 + 3$

⓭ $2x^2 = 4x - 1$

⓮ $2x^2 - 6x + 4 = 0$

⓯ $x = 5x^2 - 4$

⓰ $0 = x^2 + 5x - 3$

⓱ $-4x - 6 = -x^2$

⓲ $-4 = 4x^2 + 18x$

❶ $-4 = -3x^2 - 8x$

❷ $-7 = 4x^2 - 16x$

❸ $-8 = -2x^2 - 2x$

❹ $x^2 + 1 = -16x$

❺ $3x^2 - 15x + 3 = 0$

❻ $3x^2 + 3 = -11x$

❼ $-x^2 + 2 = 6x$

❽ $0 = x^2 + 8x - 9$

❾ $4x^2 - 2x - 1 = 0$

❿ $-4x^2 = 16x - 4$

⓫ $x^2 = -2x + 5$

⓬ $3x^2 - 7 = -x$

⓭ $14x = 4x^2 + 6$

⓮ $-2x^2 - 3 = -9x$

⓯ $4x^2 - 16x = -6$

⓰ $-3x^2 + 5x = -5$

⓱ $17x + 9 = 2x^2$

⓲ $-3 = 5x^2 + 12x$

1. $-3x^2 + 10x = -3$
2. $5x^2 + 3 = 14x$
3. $0 = 3x^2 - 10x + 7$
4. $18x = -3x^2 - 2$
5. $5x^2 + 10x - 5 = 0$
6. $3x^2 - 3 = -16x$
7. $-16x + 1 = -4x^2$
8. $-3x^2 - 5x + 8 = 0$
9. $14x = -5x^2 - 8$
10. $x^2 = -20x - 6$
11. $2x^2 - 6 = -10x$
12. $2x^2 + 17x - 1 = 0$
13. $-3x^2 = 12x - 3$
14. $4x^2 + 7 = 12x$
15. $-2x^2 + x + 5 = 0$
16. $2x^2 = 5x + 1$
17. $1 = -4x^2 + 14x$
18. $0 = 5x^2 + 15x + 9$

❶ $11\,x + 3 = -2\,x^2$

❷ $4\,x^2 + 2\,x = 2$

❸ $-5 = -2\,x^2 + 9\,x$

❹ $16\,x = 4\,x^2 - 7$

❺ $-x^2 + 1 = 18\,x$

❻ $3\,x^2 - 18\,x + 1 = 0$

❼ $2\,x^2 - 7 = 4\,x$

❽ $-3 = 2\,x^2 - 11\,x$

❾ $-2\,x^2 + 4 = x$

❿ $-4\,x^2 = 14\,x - 8$

⓫ $-x^2 + 2 = -14\,x$

⓬ $-4\,x^2 = -14\,x + 8$

⓭ $2 = 5\,x^2 + 9\,x$

⓮ $0 = -x^2 - 16\,x + 3$

⓯ $0 = -2\,x^2 - 8\,x + 5$

⓰ $2\,x^2 + 8 = -18\,x$

⓱ $0 = -x^2 + 20\,x + 8$

⓲ $10\,x + 8 = 2\,x^2$

❶ $-14\ x = -3\ x^2 - 3$

❷ $4\ x^2 = -10\ x - 6$

❸ $-5\ x^2 - 12\ x + 7 = 0$

❹ $-5\ x^2 - 6\ x = -9$

❺ $x^2 - 19\ x = 2$

❻ $-3\ x^2 + 7\ x + 6 = 0$

❼ $4\ x^2 + 7\ x - 3 = 0$

❽ $-2\ x^2 + 13\ x = 2$

❾ $5\ x^2 - 11\ x = -4$

❿ $3\ x + 5 = 3\ x^2$

⓫ $-5\ x - 1 = 2\ x^2$

⓬ $17\ x = -4\ x^2 - 4$

⓭ $-17\ x = -5\ x^2 + 4$

⓮ $-x + 6 = 3\ x^2$

⓯ $-x^2 = 8\ x + 5$

⓰ $0 = 3\ x^2 + 17\ x - 3$

⓱ $4\ x^2 - 13\ x + 5 = 0$

⓲ $7 = -x^2 + 8\ x$

❶ $3x^2 = -18x - 1$

❷ $-x^2 + 8x = 4$

❸ $5x^2 = 18x - 3$

❹ $4x^2 - 14x = 9$

❺ $2x + 5 = 4x^2$

❻ $12x - 6 = 3x^2$

❼ $0 = -2x^2 + 16x + 6$

❽ $-2x^2 + 8 = -4x$

❾ $-3x^2 = 4x - 2$

❿ $5 = -x^2 + 8x$

⓫ $18x - 6 = -2x^2$

⓬ $0 = 5x^2 - 3x - 3$

⓭ $5x^2 = -14x + 7$

⓮ $4x^2 + 14x + 5 = 0$

⓯ $2x^2 + 5 = 18x$

⓰ $10x = x^2 - 7$

⓱ $2x^2 = -16x + 4$

⓲ $0 = 5x^2 - 15x - 1$

❶ $-4x^2 = -5x - 9$

❷ $-8 = 2x^2 + 16x$

❸ $-x^2 - 5x - 6 = 0$

❹ $8x - 9 = -5x^2$

❺ $3x^2 + 16x = 4$

❻ $18x + 8 = 2x^2$

❼ $-5x + 2 = 2x^2$

❽ $x^2 = -7x + 9$

❾ $-6x + 6 = 2x^2$

❿ $3x^2 = x + 4$

⓫ $4x^2 - 9x + 3 = 0$

⓬ $3x^2 - 14x = 3$

⓭ $-5x^2 + 9 = -18x$

⓮ $-x^2 + 2 = -5x$

⓯ $0 = -x^2 + 11x - 8$

⓰ $4x^2 = -5x - 1$

⓱ $5x^2 + 3x - 4 = 0$

⓲ $x^2 + 6 = -9x$

❶ $-15x + 9 = -2x^2$

❷ $14x = -4x^2 + 4$

❸ $0 = -3x^2 - 15x + 3$

❹ $-5x^2 + 5 = 16x$

❺ $8x = -3x^2 + 7$

❻ $0 = -2x^2 + 20x - 2$

❼ $4x^2 + 2x - 5 = 0$

❽ $16x + 2 = -2x^2$

❾ $-x^2 - 12x + 3 = 0$

❿ $x + 4 = 4x^2$

⓫ $-4x^2 + 5 = -8x$

⓬ $-3x^2 - 12x = -6$

⓭ $0 = 3x^2 - 5x + 2$

⓮ $-x^2 + 12x - 2 = 0$

⓯ $3x^2 - 20x + 5 = 0$

⓰ $-5x^2 + 17x = -4$

⓱ $16x - 9 = 5x^2$

⓲ $-18x + 3 = 5x^2$

Chapter 6: Cross Multiplying

Proportionalities are very common. For example, suppose that you know that you can travel 360 miles with a full tank of gas. You have a quarter tank of gas left, and want to know how far you can travel before you must fill up. One way to figure this out is to setup a proportionality. The unknown, x, is how far you can travel on a quarter tank of gas. As a fraction, a quarter tank is represented as 1/4. The proportionality is x/360 = 1/4. That is, x is to 360 miles as a quarter tank is to a full tank. Algebraic proportionalities can be solved by cross multiplying.

A general algebraic proportionality may have unknowns in one or more numerators and/or denominators. Here are a few examples:

$$\frac{4}{x}=\frac{2}{3} \quad , \quad \frac{x+2}{3}=\frac{3-x}{4} \quad , \quad \frac{12}{x+1}=\frac{x-2}{2}$$

Equations of this form can be solved by cross multiplying. This means to multiply both sides of the equation by both denominators. Here is an example:

$$\frac{2}{x}=\frac{x}{8} \quad \rightarrow \quad 2(8)=x^2$$

We multiplied both sides by x and both sides by 8. Doing so cancels both denominators. The resulting equation can be solved using the techniques from Chapters 1 thru 5. This is called cross multiplying because you can visualize it as multiplying across the diagonals:

Multiplying down to the right, the first numerator times the second denominator gives (2)(8), and multiplying up and to the right, the first denominator times the second numerator gives x^2.

Following is the strategy for how to solve the exercises of this chapter:

- First cross multiply: Multiply the numerator of the left fraction with the denominator of the right fraction, then multiply the denominator of the left fraction with the numerator of the right fraction, and set these two products equal to each other. For example, in 5/x = 2/3, the first product is (5)(3) and the second product is 2x. Setting these products equal to each other gives 15 = 2x.
- For all of the exercises of this chapter, the resulting equation will either be a linear equation that can be solved using the strategy of Chapter 1 or a quadratic equation that can be solved using a strategy from Chapters 3 thru 5.

Example 1: $\frac{2x-3}{4}=\frac{5-2x}{3}$. Cross multiplying, the numerator of the left fraction, 2x – 3, times the denominator of the right fraction, 3, equals the denominator of the left fraction, 4, times the numerator of the right fraction, 5 – 2x: 3(2x – 3) = 4(5 – 2x). Next, we distribute: 6x – 9 = 20 – 8x. Move the 9 to the right by adding 9 to both sides (since it is being subtracted) and move the 8x to the left by adding 8x to both sides (since it is also being subtracted) to separate variables from constants: 6x + 8x = 20 + 9. Collect terms: 14x = 29. Divide by the coefficient of x: x = 29/14. Check your answer by plugging this into the original equation: [2(29/14) – 3]/4 = (29/7 – 3)/4 = (8/7)/4 = 2/7 and [5 – 2(29/14)]/3 = (5 – 29/7)/3 = (6/7)/3 = 2/7.

Example 2: $\frac{2}{4-x}=\frac{3+x}{5}$. Cross multiplying, the numerator of the left fraction, 2, times the denominator of the right fraction, 5, equals the denominator of the left fraction, 4 – x, times the numerator of the right fraction, 3 + x: 2(5) = (4 – x)(3 + x). Next, we multiply this out using the foil method (described in Chapter 4): 10 = 12 + x – x^2. Move the 10 to the right side by subtracting 10 from both sides (since it is positive) in order to put all terms of the quadratic equation on the same side, and rearrange the terms in standard form: 0 = $-x^2$+ x + 2. Identify the coefficients of the quadratic equation: a = –1, b = 1, and c = 2. Plug these into the quadratic formula: $x=\frac{-1\pm\sqrt{1^2-4(-1)(2)}}{2(-1)}=\frac{-1\pm\sqrt{1+8}}{-2}=\frac{-1\pm\sqrt{9}}{-2}=\frac{-1\pm3}{-2}$. The two solutions are x = –1 and x = 2. Check your answers by plugging them into the original equation: First, 2/[4 – (–1)] = 2/5 and [3 + (–1)]/5 = 2/5, and second, 2/(4 – 2) = 1 and (3 + 2)/5 = 1.

Example 3: $\frac{2}{8x-4}=\frac{3}{6+9x}$.

2(6 + 9x) = 3(8x – 4)

12 + 18x = 24x – 12

24 = 6x

$\boxed{x = 4}$

Check: 2(6 + 9x) = 2[6 + 9(4)] = 2(6 + 36) = 2(42) = 84.

3(8x – 4) = 3[8(4) – 4] = 3(32 – 4) = 3(28) = 84. ✓

Example 4: $\frac{1}{x+2}=\frac{3+x}{4}$.

(1)(4) = (x + 2)(3 + x)

4 = x^2 + 5x + 6

x^2 + 5x + 2 = 0

a = 1, b = 5, c = 2

$$x=\frac{-5\pm\sqrt{5^2-4(1)(2)}}{2(1)}=\frac{-5\pm\sqrt{25-8}}{2}$$

$$x=\frac{-5\pm\sqrt{17}}{2}$$

$\boxed{x = -5/2 \pm \sqrt{17}/2}$

Check: $x^2 + 5x + 6 = (-5/2 \pm \sqrt{17}/2)^2 + 5(-5/2 \pm \sqrt{17}/2) + 6 = 25/4 \mp 5\sqrt{17}/2 + 17/4 - 25/2 \pm 5\sqrt{17}/2 + 6 = 4$. ✓

❶ $\frac{8}{x-9} = \frac{x+3}{2}$

❷ $\frac{4x-8}{4x-1} = \frac{-3}{-8}$

❸ $\frac{3x+3}{-1} = \frac{9}{-4x-5}$

❹ $\frac{-5x+5}{-1} = \frac{-4x-1}{4}$

❺ $\frac{9x-4}{-5} = \frac{-3x-8}{-9}$

❻ $\frac{-9x+2}{x-4} = \frac{-2}{3}$

❼ $\frac{9x+2}{4} = \frac{-3}{6x-5}$

❽ $\frac{4}{-5x-8} = \frac{-1}{-x+4}$

❾ $\frac{-6}{-4x-8} = \frac{-7}{6x-1}$

❿ $\frac{9x+3}{-2} = \frac{8x+5}{-2}$

⓫ $\frac{-2x-7}{-1} = \frac{-5}{4x+4}$

⓬ $\frac{8x-4}{4} = \frac{6x+8}{8}$

⓭ $\frac{9x-3}{-9x} = \frac{4}{3}$

⓮ $\frac{3x-4}{-4x-4} = \frac{4}{-6}$

⓯ $\frac{-9x+1}{-6} = \frac{-1}{8x+4}$

❶ $\frac{3}{2x-5} = \frac{-1}{-4x-7}$

❷ $\frac{4}{3x+1} = \frac{3x+8}{-3}$

❸ $\frac{-5}{-2x-4} = \frac{2x+5}{2}$

❹ $\frac{6x}{2x-1} = \frac{-3}{-7}$

❺ $\frac{-2}{-9x-2} = \frac{-1}{3x+2}$

❻ $\frac{4x-4}{6x+2} = \frac{7}{2}$

❼ $\frac{x-5}{7x-6} = \frac{1}{8}$

❽ $\frac{-2x-9}{-9} = \frac{1}{x}$

❾ $\frac{-9x+3}{-4} = \frac{1}{-5x+5}$

❿ $\frac{4x+7}{-7} = \frac{-2}{x+6}$

⓫ $\frac{-4x-2}{5x+2} = -1$

⓬ $\frac{-4x}{-6} = \frac{7x}{-7}$

⓭ $\frac{-4}{-3x+6} = \frac{x-9}{-8}$

⓮ $\frac{-6x+8}{7x+5} = \frac{1}{9}$

⓯ $\frac{-9x-8}{5x+7} = -9$

❶ $\frac{4x-6}{-8} = \frac{1}{-6x+9}$

❷ $\frac{5x-4}{-2} = \frac{-6x-5}{-7}$

❸ $\frac{3x+5}{2x-4} = \frac{-5}{4}$

❹ $\frac{-2x+7}{-8x-2} = \frac{-4}{-1}$

❺ $\frac{8}{-x-1} = \frac{x+1}{-5}$

❻ $\frac{-x}{4x+6} = \frac{8}{9}$

❼ $\frac{8x+7}{-6x-1} = \frac{6}{3}$

❽ $\frac{-5}{7x+4} = \frac{7x-8}{7}$

❾ $\frac{7}{4x+7} = \frac{7}{-6x-9}$

❿ $\frac{2x+6}{-6} = \frac{-9}{x-1}$

⓫ $\frac{-8x+6}{9x+4} = \frac{-1}{-4}$

⓬ $\frac{1}{4x+7} = \frac{3x}{9}$

⓭ $\frac{-9x+6}{x+5} = \frac{-3}{8}$

⓮ $\frac{-7}{-3x-2} = \frac{9}{6x-6}$

⓯ $\frac{-3}{5x-8} = \frac{2}{-7x+1}$

❶ $\frac{-1}{-7x-1} = \frac{-3x+4}{8}$

❷ $\frac{1}{6x+5} = \frac{-9x-6}{-3}$

❸ $\frac{9x+6}{-2} = \frac{2}{5x}$

❹ $\frac{-5x+2}{1} = \frac{-3}{-4x-4}$

❺ $\frac{-8x-8}{-4} = \frac{9x+5}{4}$

❻ $\frac{3x-5}{-3} = \frac{8}{3x+9}$

❼ $\frac{3}{-7x-1} = \frac{6x+7}{2}$

❽ $\frac{2x-5}{-7} = \frac{-1}{2x+1}$

❾ $\frac{-7x+2}{4} = \frac{5x-4}{7}$

❿ $\frac{-x+3}{-5x+8} = \frac{-7}{-6}$

⓫ $\frac{-3x+6}{x-5} = \frac{-1}{-8}$

⓬ $\frac{5}{-7x-9} = \frac{7x-1}{4}$

⓭ $\frac{-4}{-5x-2} = \frac{-6}{-5x+6}$

⓮ $\frac{-5x+6}{-1} = \frac{-1}{8x-7}$

⓯ $\frac{-3x-2}{-3} = \frac{-3}{-x-6}$

❶ $\frac{-6x-6}{7x+2} = \frac{-6}{3}$

❷ $\frac{2x+5}{-2} = \frac{-9}{2x+1}$

❸ $\frac{x+9}{-3x+2} = \frac{7}{6}$

❹ $2x-4 = \frac{-4}{8x-6}$

❺ $\frac{6}{x+2} = \frac{-x-2}{-2}$

❻ $\frac{-3}{x+9} = \frac{-3x-9}{-8}$

❼ $\frac{x+2}{3x-5} = \frac{-1}{5}$

❽ $\frac{8}{-3x+3} = \frac{9}{8x-1}$

❾ $\frac{-2x+3}{2} = \frac{4}{-4x+5}$

❿ $\frac{-4}{-9x-8} = \frac{3x-2}{-3}$

⓫ $\frac{-2}{x-2} = \frac{-3x+7}{1}$

⓬ $\frac{1}{9x-9} = \frac{-2x+4}{-1}$

⓭ $\frac{-5x+2}{3x+7} = \frac{-1}{3}$

⓮ $\frac{-4x-7}{9} = \frac{-9x+3}{-6}$

⓯ $\frac{8}{4x+6} = \frac{2}{2x-9}$

❶ $$\frac{-2x+4}{-1} = \frac{7}{-6x-4}$$

❷ $$\frac{2}{3x+6} = \frac{3x-1}{-4}$$

❸ $$\frac{-2x+3}{9x-2} = \frac{-3}{-4}$$

❹ $$\frac{-2x+2}{-1} = \frac{-1}{6x-1}$$

❺ $$\frac{-x-3}{-4} = \frac{6}{3x+5}$$

❻ $$\frac{4x-9}{-4} = \frac{x-1}{6}$$

❼ $$x+7 = \frac{-4}{-2x+4}$$

❽ $$\frac{-2x+1}{-9x-6} = \frac{5}{-7}$$

❾ $$\frac{-4}{-x+1} = \frac{-x-8}{-9}$$

❿ $$\frac{-7x-6}{7x-3} = \frac{7}{-5}$$

⓫ $$\frac{5x+3}{4x-4} = \frac{1}{2}$$

⓬ $$\frac{x+4}{-4x-6} = -1$$

⓭ $$\frac{-5}{-9x+3} = \frac{3}{-5x+4}$$

⓮ $$\frac{-7x-8}{-6} = \frac{5x+6}{-1}$$

⓯ $$\frac{4}{5x-1} = \frac{-8x}{-1}$$

❶ $\frac{-2x+2}{2} = \frac{-7}{2x-4}$

❷ $\frac{1}{-7x-4} = \frac{7x+7}{-2}$

❸ $\frac{6x+1}{-7x-3} = \frac{5}{-1}$

❹ $\frac{2}{-6x-9} = \frac{x+7}{6}$

❺ $\frac{-1}{-7x+3} = \frac{x+1}{7}$

❻ $\frac{8}{-6x+3} = \frac{-3x-9}{-6}$

❼ $\frac{-9x+1}{6} = \frac{5x-3}{3}$

❽ $\frac{-2}{-2x-2} = \frac{-2x+4}{2}$

❾ $\frac{-2}{4x-7} = \frac{8}{x+3}$

❿ $\frac{x-3}{6} = \frac{-6}{-2x-3}$

⓫ $\frac{-9x+1}{6} = \frac{1}{9x-6}$

⓬ $\frac{2x-8}{-3} = \frac{-2}{-4x+5}$

⓭ $\frac{6x+9}{-2x-3} = \frac{1}{2}$

⓮ $\frac{3}{3x-5} = \frac{-9x+2}{2}$

⓯ $\frac{4x+5}{-2} = \frac{-3}{-2x+3}$

❶ $\frac{-9}{-x+1} = 6x+3$

❷ $\frac{-2}{3x-9} = \frac{4x-9}{-3}$

❸ $\frac{9x-2}{-2} = \frac{2}{7x-6}$

❹ $\frac{1}{-3x+1} = \frac{-9x+3}{8}$

❺ $\frac{4x}{-7x-9} = \frac{-1}{8}$

❻ $\frac{-x+3}{-6} = \frac{-3x+4}{2}$

❼ $\frac{-5}{2x} = \frac{7}{9x-1}$

❽ $\frac{4x+6}{-6x-1} = \frac{-5}{-9}$

❾ $\frac{-4x-4}{9x} = \frac{8}{-5}$

❿ $\frac{-x+6}{-6} = \frac{-2}{2x}$

⓫ $\frac{3}{-8x-5} = \frac{-6x}{-1}$

⓬ $\frac{3}{9x+8} = \frac{-5}{-3x-4}$

⓭ $\frac{-6}{-2x+8} = \frac{6}{-2x+9}$

⓮ $\frac{-4}{-3x+3} = \frac{2x+6}{-2}$

⓯ $\frac{3}{6x+8} = \frac{9}{-5x-3}$

❶ $\frac{2x+6}{-5} = \frac{-6}{x-7}$

❷ $\frac{8x-5}{-2x-2} = \frac{-2}{-6}$

❸ $\frac{-1}{x} = \frac{-8x-5}{1}$

❹ $\frac{-2}{9x-3} = \frac{x+3}{4}$

❺ $\frac{6x+5}{-4} = \frac{2x+1}{-1}$

❻ $\frac{-4x-9}{-3x-4} = \frac{-9}{-1}$

❼ $\frac{2x-5}{-4} = \frac{-9}{x-3}$

❽ $\frac{-3x-4}{4x-7} = \frac{-5}{-9}$

❾ $\frac{-1}{8x} = \frac{-6}{9x+4}$

❿ $\frac{3x-8}{-1} = \frac{4}{7x-8}$

⓫ $\frac{2}{-3x+2} = \frac{-x}{3}$

⓬ $\frac{1}{4x+3} = 6x+7$

⓭ $\frac{4x+2}{-6} = \frac{1}{-6x-8}$

⓮ $\frac{-3x+8}{2} = \frac{2}{-x+3}$

⓯ $\frac{-9}{-4x+6} = \frac{7}{7x-6}$

❶ $\frac{-3x-7}{-5} = \frac{2}{x+3}$

❷ $\frac{2x-1}{-7} = \frac{-x-6}{8}$

❸ $\frac{-1}{2x-1} = \frac{-6x+8}{5}$

❹ $\frac{5x+4}{-5x+4} = \frac{-9}{-7}$

❺ $\frac{-2}{3x-2} = \frac{-2}{-4x-1}$

❻ $\frac{-9}{-3x+8} = \frac{-2}{-4x+8}$

❼ $\frac{4x+6}{-7x+7} = \frac{5}{3}$

❽ $\frac{-8x+8}{5} = \frac{1}{-2x+3}$

❾ $\frac{-2}{-6x+9} = \frac{5}{-3x-8}$

❿ $\frac{-9}{-x+1} = 6x+3$

⓫ $\frac{4x+8}{7} = \frac{-8x-7}{-6}$

⓬ $\frac{6x+8}{-9} = \frac{1}{6x+2}$

⓭ $\frac{x+6}{-2} = \frac{-3}{3x+8}$

⓮ $\frac{x+8}{-8} = \frac{7}{x-9}$

⓯ $\frac{-9x-7}{x-1} = \frac{3}{2}$

❶ $\dfrac{3x-1}{-9} = \dfrac{7x-7}{2}$

❷ $\dfrac{-4x}{-x+8} = \dfrac{2}{-8}$

❸ $\dfrac{-3}{-3x+1} = \dfrac{-6x+8}{-3}$

❹ $\dfrac{-3x-9}{7x-6} = \dfrac{-7}{-4}$

❺ $\dfrac{-x+2}{6} = \dfrac{5}{-2x-4}$

❻ $\dfrac{-4}{-2x+3} = \dfrac{8x+4}{-4}$

❼ $\dfrac{-8}{-5x-4} = \dfrac{4x-4}{-1}$

❽ $\dfrac{5x-9}{-8} = \dfrac{7x+1}{4}$

❾ $\dfrac{-2}{3x-4} = \dfrac{6x+4}{7}$

❿ $\dfrac{-8x-8}{9x-4} = \dfrac{-8}{2}$

⓫ $\dfrac{8}{-8x} = \dfrac{-x+4}{-1}$

⓬ $\dfrac{7x-5}{-6x+8} = \dfrac{9}{-8}$

⓭ $\dfrac{-1}{-5x-3} = \dfrac{-8x+4}{8}$

⓮ $\dfrac{7}{-6x+1} = \dfrac{-x-5}{-2}$

⓯ $\dfrac{-x-6}{-7x} = 2$

❶ $\frac{-8x+6}{-7x-7} = \frac{6}{7}$

❷ $\frac{-x-4}{-9} = \frac{x+2}{-3}$

❸ $x+5 = \frac{-4}{-5x-1}$

❹ $\frac{9x-2}{6} = \frac{-6x-1}{-7}$

❺ $\frac{5x+9}{2} = \frac{8}{-5x}$

❻ $\frac{3x+8}{-2x+8} = \frac{2}{-5}$

❼ $-6x+7 = \frac{6}{-x+2}$

❽ $\frac{5}{-6x+4} = \frac{1}{x-1}$

❾ $\frac{9}{-3x} = \frac{2x}{-1}$

❿ $\frac{3}{4x+6} = \frac{-6}{7x+5}$

⓫ $\frac{1}{x+1} = 7x+5$

⓬ $\frac{-4x+7}{-4x-5} = \frac{-5}{6}$

⓭ $\frac{x+5}{8x-2} = 3$

⓮ $\frac{-4x-6}{-1} = \frac{3}{-x-5}$

⓯ $\frac{5x-9}{-4x+1} = \frac{-9}{8}$

❶ $$\frac{8}{-7x-8}=\frac{-1}{-x-4}$$

❷ $$\frac{1}{6x-4}=\frac{4}{5x+5}$$

❸ $$\frac{-5}{-x-8}=\frac{2}{4x+6}$$

❹ $$\frac{4x+4}{7}=\frac{-5}{3x-9}$$

❺ $$\frac{4}{9x+8}=\frac{3}{-7x-3}$$

❻ $$\frac{x-3}{7x+8}=\frac{4}{-1}$$

❼ $$\frac{-2}{-x+3}=\frac{-x-9}{4}$$

❽ $$\frac{6x-1}{8}=\frac{-7x+2}{8}$$

❾ $$\frac{x+9}{-4}=\frac{-9}{2x}$$

❿ $$\frac{-2}{2x}=\frac{6x}{-8}$$

⓫ $$\frac{4x}{2}=\frac{7}{3x-2}$$

⓬ $$\frac{x-3}{8x-1}=\frac{8}{-9}$$

⓭ $$\frac{x+2}{-3}=\frac{6}{-4x-2}$$

⓮ $$\frac{9}{-8x-6}=\frac{7}{-9x-1}$$

⓯ $$\frac{-2}{2x+1}=\frac{-x-5}{-1}$$

❶ $\frac{-3}{3x+3} = \frac{5x-7}{6}$

❷ $\frac{-5x-9}{-7} = \frac{x+7}{-9}$

❸ $\frac{-6x-7}{-8x-7} = \frac{4}{-2}$

❹ $\frac{6}{2x+1} = 4x-4$

❺ $\frac{4}{-x-2} = \frac{2x-8}{-1}$

❻ $\frac{2x-3}{2} = \frac{1}{5x-9}$

❼ $\frac{-3x+1}{-1} = \frac{-9}{-4x}$

❽ $\frac{-7}{5x-5} = \frac{2}{-7x+9}$

❾ $\frac{3}{-2x+5} = \frac{-3x-4}{-6}$

❿ $\frac{-1}{-6x+5} = \frac{-x+3}{5}$

⓫ $\frac{1}{9x} = \frac{7x+3}{-1}$

⓬ $\frac{3}{-x-5} = \frac{x-5}{4}$

⓭ $\frac{6}{x-9} = \frac{-1}{3x-3}$

⓮ $\frac{-2}{-x-2} = \frac{-6x+4}{-2}$

⓯ $\frac{3}{6x+2} = \frac{8x-7}{-5}$

❶ $-9x = \dfrac{6}{-3x+1}$

❷ $\dfrac{7}{4x} = \dfrac{2x-4}{3}$

❸ $\dfrac{3}{-5x-8} = \dfrac{-3x-5}{2}$

❹ $\dfrac{2x-9}{-8} = \dfrac{-5}{2x-1}$

❺ $\dfrac{-2x}{-4} = \dfrac{1}{3x}$

❻ $\dfrac{2x+3}{-x-4} = \dfrac{-1}{8}$

❼ $\dfrac{5}{-5x+7} = \dfrac{-3}{-8x+9}$

❽ $\dfrac{4}{6x-9} = \dfrac{3x+5}{-9}$

❾ $\dfrac{-1}{x-5} = \dfrac{-x-4}{4}$

❿ $\dfrac{9x-2}{3x+4} = \dfrac{-3}{9}$

⓫ $\dfrac{2}{3x+2} = \dfrac{7x-1}{2}$

⓬ $\dfrac{-8}{4x+7} = \dfrac{5}{5x+6}$

⓭ $\dfrac{2}{-3x-4} = \dfrac{-7x-2}{-1}$

⓮ $\dfrac{-5}{2x-2} = \dfrac{3x-6}{-3}$

⓯ $\dfrac{-6x+5}{3} = 3x-4$

❶ $\frac{1}{8x+5} = \frac{4x}{-3}$

❷ $\frac{4x+7}{-4} = \frac{4}{4x-2}$

❸ $\frac{-6}{-2x} = \frac{x}{2}$

❹ $\frac{2}{x+7} = \frac{-1}{8x+3}$

❺ $\frac{-4x-5}{-5} = \frac{-x-1}{-7}$

❻ $\frac{-1}{-x-2} = \frac{-9x-9}{-8}$

❼ $\frac{4}{2x+2} = \frac{-4x-6}{-6}$

❽ $\frac{-2}{2x+8} = \frac{x-4}{5}$

❾ $\frac{7x-7}{-6x} = 1$

❿ $\frac{-8x}{-7} = \frac{-x-8}{7}$

⓫ $\frac{-2x-5}{-5} = \frac{5}{-2x+8}$

⓬ $\frac{6x}{-5} = \frac{x-8}{2}$

⓭ $\frac{x+2}{-5} = \frac{7}{-5x-5}$

⓮ $\frac{-1}{7x+1} = \frac{4x-4}{9}$

⓯ $\frac{3}{3x+4} = \frac{-8}{6x+8}$

❶ $\frac{-9x+6}{-3x-2} = \frac{-2}{2}$

❷ $\frac{x+2}{2x+3} = \frac{-7}{-2}$

❸ $\frac{-x-1}{3} = \frac{-2}{-2x+7}$

❹ $\frac{8}{2x-6} = \frac{x-2}{5}$

❺ $\frac{-7x-3}{4x-1} = \frac{-7}{9}$

❻ $\frac{-9x-9}{3} = \frac{3x-8}{-7}$

❼ $\frac{7x+3}{2} = \frac{-3}{-4x-4}$

❽ $\frac{-6x+3}{-1} = \frac{-7}{-2x+5}$

❾ $\frac{-2x-2}{5} = \frac{1}{6x-5}$

❿ $\frac{-9}{9x-1} = \frac{8}{-9x-6}$

⓫ $2x+5 = \frac{3}{-3x-1}$

⓬ $\frac{8}{-8x+8} = -3x+4$

⓭ $-7x-6 = \frac{8x+4}{-5}$

⓮ $\frac{-5}{-3x+5} = \frac{8}{2x-9}$

⓯ $\frac{-3}{-8x+7} = -4x+1$

❶ $\frac{x+3}{-1} = \frac{-8}{6x+6}$

❷ $\frac{-x-1}{-3x-9} = \frac{-8}{7}$

❸ $\frac{9x+3}{-8} = \frac{5x-7}{8}$

❹ $\frac{-x-2}{-5} = \frac{1}{2x+6}$

❺ $\frac{1}{2x-2} = \frac{-3x-5}{-2}$

❻ $\frac{-3x+6}{-4} = \frac{9}{-2x-6}$

❼ $\frac{9x+5}{-3x-9} = \frac{4}{-4}$

❽ $\frac{-6x-6}{-1} = \frac{6}{-2x+2}$

❾ $\frac{-8}{-9x+3} = \frac{-5}{4x-6}$

❿ $\frac{3x+2}{-5x+6} = \frac{-1}{7}$

⓫ $\frac{-x+2}{-2} = \frac{-5}{-x+9}$

⓬ $\frac{-2x-2}{3x-4} = \frac{6}{5}$

⓭ $\frac{x+1}{-9} = \frac{3x+3}{-4}$

⓮ $\frac{-2x-4}{7} = \frac{-6x+3}{-4}$

⓯ $\frac{-7x+2}{-8x-8} = \frac{-2}{-7}$

❶ $\frac{8}{-8x+4}=\frac{6}{-8x+8}$

❷ $\frac{8x-2}{3x-1}=\frac{-3}{5}$

❸ $\frac{-5x-3}{7x-8}=\frac{-1}{8}$

❹ $\frac{6x+3}{-6}=\frac{5}{6x-9}$

❺ $\frac{6x+2}{-x-9}=9$

❻ $\frac{3}{-4x+4}=\frac{-x-5}{-9}$

❼ $\frac{5x-4}{3}=\frac{-5}{-x+6}$

❽ $\frac{x+7}{6}=\frac{-1}{3x+7}$

❾ $\frac{-2x-4}{-2}=\frac{-7}{6x-4}$

❿ $\frac{8}{-9x+9}=\frac{x-6}{6}$

⓫ $\frac{-7}{-2x}=\frac{-4x+3}{-1}$

⓬ $\frac{-3x+4}{3}=\frac{6x-5}{2}$

⓭ $\frac{-2x-2}{-5}=\frac{1}{4x+8}$

⓮ $\frac{2}{-4x+5}=\frac{-7}{-7x+7}$

⓯ $\frac{2x+9}{-4}=\frac{8}{-2x-7}$

❶ $\dfrac{6x+4}{-8x+1} = \dfrac{2}{3}$

❷ $\dfrac{-9}{-4x-8} = \dfrac{-4x-7}{-1}$

❸ $\dfrac{1}{-x+6} = \dfrac{2x-8}{-2}$

❹ $\dfrac{x+2}{3} = \dfrac{1}{4x+1}$

❺ $\dfrac{-2x+7}{3} = \dfrac{3}{4x+2}$

❻ $\dfrac{x-6}{-3} = \dfrac{-8x-2}{-4}$

❼ $\dfrac{x+8}{-4x-6} = \dfrac{-3}{3}$

❽ $\dfrac{-1}{5x-2} = \dfrac{6x+2}{2}$

❾ $\dfrac{-2x-2}{4} = \dfrac{2}{-8x-3}$

❿ $\dfrac{-9x+9}{9} = \dfrac{-x+9}{-5}$

⓫ $\dfrac{2x}{7} = \dfrac{-2}{-3x+6}$

⓬ $\dfrac{-3x+5}{-1} = \dfrac{-5}{-3x}$

⓭ $\dfrac{-x+4}{8} = \dfrac{-3}{-8x}$

⓮ $\dfrac{6x-7}{-6} = \dfrac{2}{8x-1}$

⓯ $\dfrac{-3}{x+4} = \dfrac{-2x-7}{2}$

❶ $\frac{-3}{x-2} = \frac{-6x+4}{8}$

❷ $\frac{7x+6}{-9} = \frac{-6x-4}{4}$

❸ $\frac{7}{-3x-3} = \frac{-5}{-x+3}$

❹ $\frac{-1}{-6x+2} = \frac{3x+2}{9}$

❺ $\frac{1}{2x-2} = 8x+6$

❻ $\frac{-1}{8x+5} = \frac{4x+3}{-4}$

❼ $\frac{4}{-8x-8} = \frac{-x-3}{-1}$

❽ $\frac{-2}{-8x-6} = \frac{-4}{-2x+3}$

❾ $\frac{-3x+8}{-3} = \frac{3}{-4x}$

❿ $\frac{-3}{6x+3} = \frac{5x+1}{-2}$

⓫ $\frac{9x+5}{-2x-8} = \frac{1}{-4}$

⓬ $\frac{7}{-x-8} = \frac{4x+4}{5}$

⓭ $\frac{2}{4x-4} = \frac{-3x-7}{9}$

⓮ $\frac{-1}{-x+4} = \frac{-x-4}{6}$

⓯ $\frac{-3x-1}{2} = \frac{6x+7}{-7}$

❶ $\frac{-8}{-8x-8} = \frac{4}{5x+7}$

❷ $\frac{5x+6}{-x-7} = \frac{-7}{-5}$

❸ $\frac{2x+1}{5} = \frac{3}{3x+3}$

❹ $\frac{-4x-3}{-x+1} = \frac{-1}{4}$

❺ $\frac{8x-4}{-6x+8} = \frac{-5}{-4}$

❻ $5x+8 = \frac{6}{-5x-1}$

❼ $\frac{-2}{6x-4} = \frac{9}{-7x+9}$

❽ $\frac{9x}{3} = \frac{1}{8x}$

❾ $\frac{6x+2}{7x-3} = \frac{1}{-6}$

❿ $\frac{-9}{-x+7} = \frac{3x-9}{5}$

⓫ $7x = \frac{-3}{5x-3}$

⓬ $\frac{-4}{-9x-8} = \frac{1}{-8x-9}$

⓭ $\frac{x-8}{2} = \frac{9}{-2x+3}$

⓮ $\frac{-8}{-x-4} = \frac{9}{-x-9}$

⓯ $\frac{-5}{6x+7} = \frac{2x-7}{9}$

❶ $\frac{-4x-6}{8} = \frac{1}{6x-1}$

❷ $\frac{3x}{-2x+5} = \frac{-7}{-6}$

❸ $\frac{-7x+7}{-7} = \frac{-2}{9x}$

❹ $\frac{-3x-5}{-6x+2} = \frac{-8}{4}$

❺ $\frac{4x+6}{7x-5} = \frac{-3}{3}$

❻ $\frac{-2x-8}{-7x+9} = \frac{-3}{-5}$

❼ $\frac{-8x+6}{-6} = \frac{-5}{2x+5}$

❽ $\frac{-4x}{2} = \frac{9}{-3x+2}$

❾ $\frac{-x-8}{2x+1} = \frac{2}{-3}$

❿ $\frac{-4x-2}{5} = \frac{-x-6}{9}$

⓫ $\frac{-3}{-x+4} = \frac{-2x+3}{-6}$

⓬ $\frac{-3}{-x-9} = \frac{3x-3}{-3}$

⓭ $\frac{3x+4}{3x-7} = \frac{-2}{-3}$

⓮ $\frac{-7x-9}{9} = \frac{-7x+7}{8}$

⓯ $\frac{-x+6}{-9} = \frac{-5}{-2x+6}$

❶ $\frac{-6}{-x-9} = \frac{3x+3}{-8}$

❷ $\frac{5x-3}{-3x+5} = \frac{5}{6}$

❸ $\frac{-3}{5x+2} = \frac{1}{-3x+6}$

❹ $\frac{7}{-x-9} = \frac{x-9}{-9}$

❺ $\frac{3}{x+4} = \frac{-x+7}{5}$

❻ $\frac{-6}{6x-8} = \frac{-2}{9x+2}$

❼ $\frac{-2x-4}{-5} = \frac{-3x-6}{4}$

❽ $\frac{-5x+9}{-4} = \frac{-7}{8x}$

❾ $\frac{-6}{-x-5} = \frac{x-1}{4}$

❿ $\frac{1}{-4x+2} = \frac{-2x+4}{3}$

⓫ $\frac{7}{x} = \frac{-5}{3x+3}$

⓬ $\frac{-9}{-9x+3} = \frac{3x+8}{-6}$

⓭ $\frac{1}{-x+9} = \frac{2x-7}{3}$

⓮ $\frac{-2x+2}{6} = -3x+1$

⓯ $\frac{8x-6}{2} = \frac{1}{6x-3}$

Chapter 7: Systems of Linear Equations

The exercises of this chapter involve two unknowns, x and y. In general, a single equation that has two unknowns does not have a single numerical solution. For example, if $3x = 2y$, $x = 2$ and $y = 3$ solves the equation (since $3 \cdot 2 = 2 \cdot 3$), but $x = 8$ and $y = 12$ also solves the equation (since $3 \cdot 8 = 2 \cdot 12$), and an infinite number of other pairs, (x, y) also solve this same equation, $3x = 2y$. However, there can be a unique solution if there are two equations in two unknowns. For example, $x = 2$ and $y = 3$ is the only solution that satisfies the two equations $3x = 2y$ and $2x - 1 = y$. Although $x = 8$ and $y = 12$ satisfies the first equation ($3 \cdot 8 = 2 \cdot 12$), it does not satisfy the second equation (since $2 \cdot 8 - 1 = 15 \neq 12$).

All of the exercises of this chapter have 2 linear equations and 2 unknowns. The most straightforward strategy for solving such a system of linear equations is the method of substitution. The main idea is to solve for one unknown in one equation and plug (substitute) it into the other equation. Following is the strategy for solving a system of linear equations using the method of substitution.

- Solve for one variable in one equation using the strategy from Chapter 1. For example, to solve for x in the equation $4x + 8y = 12$, put all of the x terms on one side of the equation and all of the y and constant terms on the other side. In this case, we can move 8y to the right by subtracting 8y from both sides (since it is positive): $4x = 12 - 8y$. Now we divide both sides by the coefficient of x: $x = (12 - 8y)/4 = 3 - 2y$.
- Next, plug this equation into the unused equation. For example, if $4x + 8y = 12$ and $2x - 3y = 5$, having found $x = 3 - 2y$ from the first equation, we plug this in for x in the second equation: $2(3 - 2y) - 3y = 5$.
- Simplify the new equation and solve for the remaining unknown. Continuing the previous example, $2(3 - 2y) - 3y = 5$ becomes $6 - 4y - 3y = 5$ after we distribute. Next, we separate the variable (those with a y) terms from the constant terms. We can have all of the constants on the left if we subtract 5 from both sides: $1 - 4y - 3y = 0$. We can get all of the y terms on the right by adding 4y and 3y to both sides: $1 = 4y + 3y$ or $1 = 7y$. Dividing by the coefficient of y, we find that $y = 1/7$.
- Now you know one of the two variables. Plug this value into any of the previous equations that has both variables to solve for the other unknown. Continuing the previous example, we can plug $y = 1/7$ into $4x + 8y = 12$ to find that $4x + 8(1/7) = 12$. This simplifies to $4x + 8/7 = 12$. Subtracting 8/7 from both sides, $4x = 12 - 8/7 = 76/7$. Dividing both sides by 4, $x = 19/7$.

Following are examples of how to apply the method of substitution to solve a system of linear equations with two unknowns. After the examples, we will learn an alternative method for solving such a system, which is often more efficient.

Example 1: $9x - 3y = 6$ and $4x + 2y = 16$. First we solve for one variable in one equation; we choose to solve for y in the second equation. We isolate the y term by moving the 4x term to the right, which we do by subtracting 4x from both sides (since it is positive): $2y = 16 - 4x$. We solve for y by dividing both sides by its coefficient: $y = (16 - 4x)/2$. Now we distribute the 1/2: $y = 8 - 2x$. After solving for one variable in one equation, we plug it into the other equation: $9x - 3(8 - 2x) = 6$. Distributing the -3, we get $9x - 24 + 6x = 6$. Collecting the x terms, $15x - 24 = 6$. Isolate the x term by moving the 24 to the right. This is done by adding 24 to both sides (since it is negative): $15x = 30$. Divide by the coefficient of x: $x = 30/15 = 2$. To solve for y, we plug the numerical value of x into any of the previous equations that have y. We could go back to an original equation, but it is most efficient to use $y = 8 - 2x = 8 - 2(2) = 8 - 4 = 4$. The final answer is $x = 2$ and $y = 4$. We can plug these values into both of the original equations to check that they are correct: $9(2) - 3(4) = 18 - 12 = 6$ and $4(2) + 2(4) = 8 + 8 = 16$.

Example 2: $2x + 6y = 9$ and $4x - 2y = 3$.

$$2x = 9 - 6y$$
$$x = (9 - 6y)/2 = 9/2 - 3y$$
$$4(9/2 - 3y) - 2y = 3$$
$$18 - 12y - 2y = 3$$
$$18 - 14y = 3$$
$$18 = 3 + 14y$$
$$15 = 14y$$
$$15/14 = y$$
$$x = 9/2 - 3y = 9/2 - 3(15/14)$$
$$x = 9/2 - 45/14 = 9/7$$
$$\boxed{x = 9/7,\ y = 15/14}$$

Check: $2(9/7) + 6(15/14) = 18/7 + 45/7 = 63/7 = 9$. ✓

$4(9/7) - 2(15/14) = 36/7 - 15/7 = 21/7 = 3$. ✓

Example 3: $x + 2y = 6$ and $3x - y = 9$.

$$x = 6 - 2y$$
$$3(6 - 2y) - y = 9$$
$$18 - 6y - y = 9$$
$$18 - 7y = 9$$
$$18 = 9 + 7y$$
$$9 = 7y$$
$$9/7 = y$$
$$x + 2(9/7) = 6$$
$$x + 18/7 = 6$$
$$x = 6 - 18/7 = 24/7$$
$$\boxed{x = 24/7,\ y = 9/7}$$

Check: $24/7 + 2(9/7) = 24/7 + 18/7 = 42/7 = 6$. ✓

$3(24/7) - 9/7 = 72/7 - 9/7 = 63/7 = 9$. ✓

Here are a few more notes:

- After you solve for the first variable in terms of the second variable, make sure you plug this equation into the unused equation (otherwise you will go around in circles or run into a dead end). On the other hand, once you solve for one of the unknowns numerically, you may plug it into any equation that contains both unknowns.
- A system does not necessarily have a unique solution. Consider $x + y = 2$ and $2x + 2y = 4$. If you multiply the first equation by 2, you get the second. This system really has the same equation twice, rather than two distinct equations.
- Not all systems of equations have solutions. For example, there is no solution to $x = y + 1$ and $x = y + 2$ because x can't be both 1 and 2 greater than y at the same time. If you solve this system, you find that $y + 1 = y + 2$, which simplifies to $1 = 2$, which is not true!

Although it is possible for a system of equations to have no solution or to have an indefinite solution, the exercises of this chapter all have one unique solution for the two variables.

The method of substitution is the most straightforward way to solve a system of equations, but it is not always the most efficient. When you apply the method of substitution, you isolate an unknown in one equation, and then plug it into the other. Doing so, you are working with one equation first and then working with the other. An alternative approach involves working with both equations together, and so is called solving the system simultaneously.

The idea of solving a system of linear equations simultaneously is to perform algebra on both equations so that the coefficients of one variable become equal and opposite. Once this happens, one of the variables is eliminated by adding the equations together. For example, if you had $x + 2y = 4$ and $3x - 2y = 6$, if you add the equations together, you get $x + 2y + 3x - 2y = 4 + 6$, which simplifies to $4x = 10$. You see: The y terms cancelled because they had equal and opposite coefficients. It is now easy to solve for x.

When you see this for the first time, it is natural to wonder if it is legitimate to add two equations together. Consider the equation $2 + 4 = 6$ and $1 + 9 = 10$. Try adding these equations together. You obtain $2 + 4 + 1 + 9 = 6 + 10$. When you add equations together, you make a new equation by setting the sum of the left sides equal to the sum of the right sides. In this case, we find that $16 = 16$. We can also look at this symbolically. Suppose that $a = b$ and $c = d$. If you add the equations together, you get $a + c = b + d$. We can write this as $a + d = a + d$, since $c = d$ and $b = a$. If you study this, you should be convinced that adding equations together is okay.

The other question you should have on your mind is how to make the coefficients equal and opposite. As an example, consider the equations $3x - 2y = 5$ and $2x + 4y = 6$. The y terms already have the opposite sign. We could make them equal and opposite by multiplying both sides of the first equation by 2: Then you would get $6x - 4y = 10$.

It's not always so easy though, so let's try another example: $5x - 3y = 2$ and $3x + 4y = 8$. Again the y terms have opposite sign, but you can't multiply any one equation by an integer to make them equal and opposite. You could multiply one by a fraction, but it is more convenient to multiply both equations by different integers. For example, if you multiply the first equation by 4 and the second equation by 3, you get $20x - 12y = 8$ and $9x + 12y = 24$.

Let's try an example where the signs are the same: $2x + 4y = 6$ and $5x + 3y = 9$. This time, we need to multiply one of the equations by a negative number. For example, we can make the x terms equal and opposite by multiplying the first equation by 5 and the second equation by -2: $10x + 20y = 30$ and $-10x - 6y = -18$.

Notice that the numbers that we've been multiplying the equations by equal the coefficients of the terms that we are trying to make equal and opposite. For example, in $5x - 3y = 2$ and $3x + 4y = 8$, we multiplied the first equation by 4 and the second equation by 3. This works, but – like finding common denominators – sometimes you can multiply by smaller numbers. For example, consider $7x + 6y = 8$ and $5x - 9y = 2$. We could multiply the first equation by 9 and the second by 6. We could also multiply the first by 3 and the second by 2.

The most convenient numbers to multiply by are the same as if you were trying to make the lowest common denominator. If you were adding 1/6 to 1/9, the lowest common denominator would be 18, since 18 = 6(3) and 18 = 9(2). Compare with the 3 and 2 mentioned previously.

Following is the strategy for how to solve a system of two linear equations simultaneously.

- Choose which variable – x or y – to eliminate by making its coefficients equal and opposite. Sometimes, one will be a little more convenient than the other. For example, if one variable's terms are already opposite in sign, that may be more convenient. Also, think about making the lowest common denominator from the coefficients – one might be simpler. For example, if $2x + 9y = 14$ and $x - 2y = 12$, although the y terms are opposite in sign, you would have to multiply one equation by 3 and the other by 9. However, to eliminate x you only have to multiply the second equation by –2.
- Multiply each equation by a number to make equal and opposite coefficients for the desired variable. It is most convenient to think in terms of the lowest common denominator, even if you're not working with fractions. For example, suppose that you wish to eliminate x from $4x - 8y = 1$ and $-6x - 7y = 9$. The lowest common denominator that you can make from 4 and 6 is 12, so we multiply the first equation by 3 and the second equation by 2 (both numbers are positive since the signs are already opposite).
- Once the coefficients are equal and opposite, simply add both sides of the two equations together. For example, if you have $5x + 2y = 9$ and $3x - 2y = 4$, you get $5x + 2y + 3x - 2y = 9 + 4$.
- Simplify the resulting equation and solve for the remaining unknown. In the case of $5x + 2y + 3x - 2y = 9 + 4$, this reduces to $8x = 13$. Remember, one variable will disappear because its coefficients are equal and opposite (otherwise, you made a mistake). Now we divide both sides by the coefficient of x: $x = 13/8$.
- After solving for one unknown numerically, plug it into any of the previous equations that has both variables (x and y) in order to solve for the other unknown. This last step is the same for both the substitution and simultaneous equations methods.

The method of simultaneous equations is usually more efficient than the method of substitution. This is the benefit of working with both equations together. However, there is one occasion where the method of substitution is often convenient: That's when either x or y has a coefficient of 1 in one of the equations. For example, consider $x + 4y = 5$ and $8x - 2y = 10$. We could multiply the second equation by 2 and then add the equations together. Alternatively, it would be easy to solve for x in the first equation just by moving 4y to the other side. It's often an advantage to be proficient at multiple methods of solving equations when there is more than one way to solve it.

Following are a few examples of solving a system of two linear equations using the method of simultaneous equations.

Example 1: $9x - 3y = 6$ and $4x + 2y = 16$. We can multiply the first equation by 2 and the second equation by 3 to get $18x - 6y = 12$ and $12x + 6y = 48$. Adding the two equations together results in $18x + 12x = 12 + 48$, which simplifies to $30x = 60$. Divide by the coefficient of x to find that $x = 2$. Now plug this into $4x + 2y = 16$ to get $4(2) + 2y = 16$. Subtract 8 from both sides to isolate the y term: $2y = 16 - 8 = 8$. Divide both sides by the coefficient of y: $y = 8/2 = 4$. The final answer is $x = 2$ and $y = 4$. Note that this was the same first example of the method of substitution – you may wish to compare the solutions.

Example 2: $3x + 5y = 8$ and $-5x + 9y = 4$.

$$5(3x + 5y = 8)$$
$$3(-5x + 9y = 4)$$

$$15x + 25y = 40$$
$$-15x + 27y = 12$$

$$52y = 52$$
$$y = 1$$
$$3x + 5(1) = 8$$
$$3x = 8 - 5 = 3$$
$$x = 1$$
$$\boxed{x = 1, y = 1}$$

Check: $3(1) + 5(1) = 3 + 5 = 8$. ✓
$-5(1) + 9(1) = -5 + 9 = 4$. ✓

Example 3: $9x + 8y = 2$ and $-7x + 6y = -9$.

$$3(9x + 8y = 2)$$
$$-4(-7x + 6y = -9)$$

$$27x + 24y = 6$$
$$28x - 24y = 36$$

$$55x = 42$$
$$x = 42/55$$
$$9(42/55) + 8y = 2$$
$$378/55 + 8y = 2$$
$$8y = 2 - 378/55 = -268/55$$
$$y = -67/110$$
$$\boxed{x = 42/55, y = -67/110}$$

Check: $9(42/55) + 8(-67/110)$
$= 378/55 - 268/55 = 110/55 = 2$. ✓
$-7(42/55) + 6(-67/110)$
$-294/55 - 201/55 = -495/55 = -9$. ✓

Here is another comment:

- Make sure that you multiply all of the terms of an equation by the same factor. Also, make sure that you add all of the terms from both sides of the two equations. One of the common mistakes that students make is to leave something out.

❶

$5x - 6y = -3$

$2x - 6y = -7$

❷

$6x - 7y = 2$

$5x + 8y = -4$

❸

$5x + 3y = 3$

$-2x - y = 6$

❹

$-3x - 4y = -2$

$4x - 7y = -6$

❺

$-3x + 6y = 5$

$-x + 9y = 7$

❻

$-2x + 4y = -2$

$2x + 5y = -6$

❼

$2x - 6y = -1$

$-5x + 2y = 2$

❽

$x - 4y = 7$

$8x + y = -3$

❾

$-6x - 9y = -1$

$-6x - 6y = 4$

❿

$5x - 8y = 7$

$3x - 9y = 8$

⓫

$x - 8y = 3$

$-8x + 3y = -4$

⓬

$2x - 3y = 6$

$-8x - 8y = 7$

❶
$7x + 4y = -5$
$-7x - 8y = 7$

❷
$-3x - 3y = 9$
$-x - 3y = -6$

❸
$-3x + 2y = 8$
$-9x - 4y = 4$

❹
$x + 7y = -9$
$-7x - 4y = -2$

❺
$-8x - 4y = 5$
$5x - 2y = -6$

❻
$2x + 8y = -4$
$7x - 3y = 1$

❼
$-9x + 9y = 7$
$2x + y = -7$

❽
$-5x + 9y = 2$
$-8x + 5y = 4$

❾
$8x - 9y = -9$
$-3x + 6y = -2$

❿
$7x + 4y = -2$
$7x - 7y = 6$

⓫
$7x - y = 9$
$-4x + 2y = -2$

⓬
$8x - 6y = 6$
$-x - 3y = 8$

❶

$2x - 5y = -9$

$5x - 8y = -3$

❷

$-7x - 3y = 1$

$7x + y = 6$

❸

$-5x + 5y = -7$

$5x + y = -5$

❹

$3x - 9y = 4$

$x + 3y = -5$

❺

$6x - 8y = 7$

$8x + 8y = 2$

❻

$7x - 8y = 5$

$-9x - y = 2$

❼

$-x - 6y = 7$

$-6x + y = 9$

❽

$7x - 9y = -1$

$-6x + 2y = -2$

❾

$x - 2y = -6$

$4x - 5y = 1$

❿

$5x + 5y = -5$

$-8x + 2y = -6$

⓫

$-2x - 7y = -3$

$-9x + 7y = 3$

⓬

$6x - 3y = 7$

$4x + 2y = -4$

❶

$x - 5y = -4$

$-5x + 4y = 3$

❷

$2x + 9y = 2$

$9x + y = 6$

❸

$-5x - 4y = 9$

$-9x + 9y = -8$

❹

$-7x - 5y = 3$

$6x + 7y = 1$

❺

$-x + 9y = 5$

$6x + y = -9$

❻

$-x - 7y = 2$

$2x - 2y = -2$

❼

$-5x - 9y = 2$

$4x - 8y = 7$

❽

$x - 8y = -3$

$5x - 8y = -4$

❾

$-4x + 6y = -2$

$4x + 6y = 5$

❿

$8x - 8y = -4$

$4x - 5y = 5$

⓫

$-8x + 5y = -7$

$7x - 5y = 8$

⓬

$-9x + 4y = 7$

$9x - 9y = 7$

❶
$8x + 6y = 6$
$6x - 9y = 6$

❷
$-8x + 2y = -8$
$4x + 5y = 9$

❸
$7x - 2y = -3$
$x + 6y = 8$

❹
$7x + 8y = 4$
$-2x + 7y = -2$

❺
$-7x + 9y = -1$
$6x + 4y = -2$

❻
$-7x + 6y = 2$
$-6x - 6y = 6$

❼
$-7x + 9y = -6$
$5x - 9y = 9$

❽
$-9x - 8y = 5$
$9x - 3y = -3$

❾
$3x + 5y = 1$
$-4x - 7y = 2$

❿
$-6x - 4y = 3$
$4x - 8y = 9$

⓫
$-8x + 5y = -2$
$3x + 7y = 9$

⓬
$2x + 7y = -7$
$-6x + 5y = 4$

❶

$-7x - 5y = -9$

$-4x + 6y = -3$

❷

$x + 7y = -5$

$x - 7y = 5$

❸

$9x - 8y = 7$

$-3x + 5y = -9$

❹

$4x - 6y = 3$

$6x - 7y = -2$

❺

$-2x + 5y = 7$

$7x - 6y = -5$

❻

$x - 6y = -8$

$-8x - 2y = 9$

❼

$3x - 2y = -6$

$-9x - 3y = 4$

❽

$-4x + 8y = 6$

$-2x + 5y = -7$

❾

$9x + 7y = -6$

$x + 7y = -4$

❿

$-3x + 4y = 4$

$-4x - 8y = -7$

⓫

$5x + 5y = 7$

$-x + 6y = 9$

⓬

$-7x + 5y = 9$

$-3x + 4y = 3$

❶
$-8x + 8y = 4$
$-5x + 7y = 9$

❷
$-6x + 8y = -3$
$-7x - y = 8$

❸
$8x - 8y = -3$
$x - 5y = 4$

❹
$3x + 3y = -7$
$-6x + 2y = -3$

❺
$-7x + 2y = -5$
$-x - y = 5$

❻
$8x - 7y = 1$
$3x - 9y = -1$

❼
$8x - 8y = 6$
$5x - 4y = -3$

❽
$-x - y = -4$
$4x - 7y = 1$

❾
$3x - 9y = 9$
$-9x + 3y = -9$

❿
$-5x + 5y = 2$
$6x + 4y = -3$

⓫
$-9x + 6y = -2$
$9x + 2y = 7$

⓬
$9x + 7y = 9$
$-4x + 8y = -2$

❶

$6x - 9y = -4$

$6x - 7y = 9$

❷

$x + 6y = -8$

$6x - 2y = -6$

❸

$-x - 4y = -1$

$-2x - 4y = 4$

❹

$-3x - 9y = -9$

$6x + 3y = -8$

❺

$-3x + 4y = 3$

$-x + 6y = -2$

❻

$-3x + 7y = -4$

$-5x - 3y = -5$

❼

$2x - y = 8$

$-3x + 2y = -9$

❽

$x - 7y = -3$

$-x + 6y = 4$

❾

$9x - 6y = 6$

$3x - y = 5$

❿

$8x - 9y = -9$

$9x - 6y = -1$

⓫

$-6x - 9y = -6$

$9x + 4y = -3$

⓬

$-6x + 6y = -9$

$7x + 3y = -8$

❶

$2x + 4y = 9$

$-2x + 7y = 7$

❷

$4x + y = 9$

$6x + 4y = 5$

❸

$6x - 6y = -9$

$-5x + 4y = -8$

❹

$-7x + 7y = 9$

$5x + 8y = 7$

❺

$x + 2y = 2$

$-5x - 3y = -2$

❻

$-7x - 8y = 7$

$-5x - 2y = 9$

❼

$-5x - 7y = -3$

$-6x + 8y = 6$

❽

$-6x - 6y = 8$

$9x + 6y = 1$

❾

$-6x + 9y = -9$

$6x - 2y = 8$

❿

$5x + y = -6$

$-3x + 3y = -9$

⓫

$4x - 7y = 5$

$-9x + 7y = -4$

⓬

$-4x - y = -9$

$x - 7y = -3$

❶

$x + 5y = 6$

$8x - 4y = 8$

❷

$6x - 7y = 3$

$5x + y = -7$

❸

$-3x + 4y = -7$

$-4x + 6y = -5$

❹

$7x - 9y = -9$

$-2x - y = 4$

❺

$-6x - y = 2$

$7x - 2y = 7$

❻

$-6x + y = -6$

$-5x - 4y = 1$

❼

$-6x - 5y = 3$

$x + 8y = -4$

❽

$2x - 7y = 4$

$2x - 3y = -9$

❾

$2x + 5y = -2$

$-2x - 6y = 3$

❿

$x + y = 3$

$2x + 8y = -3$

⓫

$-7x + 6y = 7$

$2x - 4y = 1$

⓬

$2x + 4y = 3$

$3x - 5y = 5$

❶
$x - 4y = -8$
$9x - 8y = -3$

❷
$9x + 2y = -8$
$8x - 8y = 6$

❸
$-4x - 9y = -2$
$-3x - 7y = 6$

❹
$-7x + 4y = -7$
$-6x + 2y = -1$

❺
$-9x - 3y = -7$
$9x + 7y = -4$

❻
$3x + 5y = 6$
$2x - 8y = -7$

❼
$-2x + 6y = -8$
$-9x - y = 3$

❽
$-6x - 6y = 6$
$-3x + 9y = -2$

❾
$-4x - 3y = 7$
$5x - 7y = 9$

❿
$-5x - 7y = -6$
$-7x + 8y = -7$

⓫
$5x + y = -6$
$5x + 2y = -5$

⓬
$-3x + 8y = 3$
$-3x - 6y = -5$

❶

$-2x - 3y = 7$

$-8x - 9y = 7$

❷

$-7x - 9y = 5$

$-4x + 8y = -5$

❸

$-5x - 8y = 5$

$-7x - 7y = -2$

❹

$4x - 9y = -9$

$-7x + 5y = -5$

❺

$4x - 6y = 9$

$-3x + y = -2$

❻

$-6x + 3y = 2$

$-6x - 3y = 6$

❼

$3x + y = 5$

$-8x + 8y = 7$

❽

$-6x + 3y = 4$

$5x + 3y = 2$

❾

$-5x + 8y = 8$

$8x - 9y = 7$

❿

$-6x - 8y = -2$

$-8x - 6y = -9$

⓫

$x - 6y = 8$

$2x - 4y = 6$

⓬

$-x - 9y = 2$

$9x - 6y = -1$

❶
$-x - 7y = -8$
$x + 8y = -8$

❷
$3x - y = 1$
$7x - 2y = -4$

❸
$9x + 3y = -3$
$9x - 9y = 8$

❹
$3x + 7y = -9$
$-6x - 6y = -8$

❺
$7x + y = -8$
$5x - y = -2$

❻
$2x + 3y = -7$
$x - 5y = 4$

❼
$-x - 2y = -7$
$6x + 9y = -7$

❽
$x + 2y = -7$
$x - y = 3$

❾
$-2x - y = 8$
$9x - y = -9$

❿
$3x - 8y = -9$
$2x + 3y = -4$

⓫
$-6x + 5y = -8$
$-4x + 4y = -4$

⓬
$6x + 5y = -2$
$-5x - 5y = 6$

❶

$2x - 6y = 8$

$-2x - 6y = -6$

❷

$3x - y = 8$

$-8x + y = -5$

❸

$-6x - 4y = -4$

$-x - 6y = 5$

❹

$4x + 2y = 7$

$9x - 7y = -7$

❺

$3x - 9y = -9$

$-4x + 4y = 7$

❻

$2x + 4y = 6$

$-7x - 9y = -9$

❼

$8x - y = 2$

$4x + 9y = -6$

❽

$6x + 2y = -9$

$3x - 7y = 2$

❾

$x - 7y = -5$

$9x - 8y = -2$

❿

$x + 5y = 8$

$-x + 2y = -7$

⓫

$5x - 3y = -6$

$-8x + 2y = 6$

⓬

$-6x + 6y = -6$

$-x - y = 8$

❶
$-7x + 3y = -3$
$6x - y = 8$

❷
$5x + 3y = 5$
$-2x + 9y = 8$

❸
$-9x - 7y = -6$
$-2x - 9y = 4$

❹
$x + 6y = -1$
$-7x + y = 8$

❺
$-7x - 8y = -6$
$-9x - 5y = -5$

❻
$-3x - 9y = -6$
$-4x - 8y = -4$

❼
$-6x - y = -6$
$7x - 5y = -5$

❽
$-9x + 4y = 3$
$6x - 5y = 5$

❾
$2x + 7y = -8$
$2x - 2y = 6$

❿
$-3x + 9y = -6$
$4x - 2y = -3$

⓫
$-x - 4y = 4$
$-9x + 9y = 1$

⓬
$9x + 4y = 7$
$2x - 6y = 3$

❶

$-2x + 9y = -4$

$-5x - 2y = 6$

❷

$-7x - 3y = -1$

$-6x - 4y = 3$

❸

$2x + 3y = -5$

$-6x + 5y = -5$

❹

$-6x + 3y = -3$

$-8x + 9y = 9$

❺

$8x + 4y = 1$

$-5x - 8y = 3$

❻

$8x + 4y = 1$

$7x - 3y = 9$

❼

$-5x - 7y = 6$

$-x + 6y = -6$

❽

$6x + 9y = 7$

$4x - 9y = 5$

❾

$4x - 2y = -9$

$-5x - 7y = 7$

❿

$9x - 9y = 1$

$-7x - 5y = 4$

⓫

$6x + 5y = -6$

$-8x + 9y = 1$

⓬

$-4x + 4y = 5$

$8x + y = -6$

❶
$-5x + 8y = -1$
$-9x + 5y = -3$

❷
$3x - 9y = 5$
$9x - 4y = -8$

❸
$-7x - 9y = 2$
$9x - 4y = 1$

❹
$-8x - 4y = 5$
$9x + 6y = 7$

❺
$-5x + 4y = 1$
$-4x - 6y = 5$

❻
$-2x + 3y = 9$
$-2x + 4y = -3$

❼
$-8x + 6y = 4$
$2x + 9y = 9$

❽
$x - 2y = 1$
$-3x + 9y = -7$

❾
$-x - 6y = -8$
$-4x + 9y = 6$

❿
$-9x + 6y = 5$
$-7x + 8y = -1$

⓫
$9x + 9y = 1$
$-8x + 9y = -1$

⓬
$9x + 6y = 4$
$4x - 6y = -2$

❶

$x + 5y = -1$

$-x + 3y = -6$

❷

$2x + 6y = -7$

$7x + 3y = -2$

❸

$-5x - 3y = 3$

$-2x + 5y = -8$

❹

$-x - 3y = -7$

$-4x - 5y = -5$

❺

$-x + 9y = 7$

$-8x + 2y = 3$

❻

$9x + 4y = -7$

$-9x - 9y = 8$

❼

$2x + 9y = 1$

$-7x - 9y = -9$

❽

$x + 2y = 4$

$-4x + 4y = 6$

❾

$-9x + 2y = 7$

$-2x + 8y = -6$

❿

$7x - 7y = -7$

$4x - 2y = -7$

⓫

$-3x + 8y = 6$

$8x + 8y = 8$

⓬

$-6x - 3y = 8$

$6x - 4y = 2$

❶

$8x + 4y = -3$

$-8x + 4y = 5$

❷

$8x - 9y = -7$

$-9x + 7y = -7$

❸

$-4x + 2y = 2$

$-4x - 6y = 2$

❹

$-9x - 7y = 5$

$6x - 2y = 2$

❺

$7x - 8y = 4$

$-8x - 2y = 5$

❻

$-4x + 2y = -9$

$4x + y = 2$

❼

$-6x + y = -9$

$-8x + 7y = -1$

❽

$4x + y = -8$

$2x - 6y = -1$

❾

$3x + 3y = 8$

$6x - 6y = -5$

❿

$2x - 7y = -9$

$x + 7y = -8$

⓫

$-6x + 2y = 8$

$-x - y = -2$

⓬

$6x + 3y = -5$

$8x - y = -9$

❶

$-5x - y = -4$

$-9x + 6y = -3$

❷

$7x - 8y = -3$

$-9x - 2y = 6$

❸

$4x + 2y = -1$

$6x + 6y = 9$

❹

$3x + 7y = -6$

$-3x + 2y = -4$

❺

$-6x + 7y = 5$

$-3x + 5y = -8$

❻

$-5x - 4y = -7$

$2x + y = 3$

❼

$9x - 9y = -8$

$-9x + 4y = -6$

❽

$8x + 5y = -2$

$6x - 4y = 1$

❾

$7x + y = 4$

$-4x + 8y = 2$

❿

$4x - 9y = 7$

$3x - 4y = 3$

⓫

$2x + 9y = 6$

$9x + 5y = 1$

⓬

$3x + 2y = 7$

$-3x + y = 4$

❶

$5x + y = 6$

$8x - 2y = 5$

❷

$-x - 7y = 2$

$-4x - 2y = -3$

❸

$-6x + 8y = 7$

$8x - 5y = -8$

❹

$-6x + 4y = 6$

$9x - 9y = 5$

❺

$6x - 5y = 6$

$2x + 9y = -5$

❻

$-7x + 3y = -9$

$6x - 2y = 5$

❼

$-3x - 4y = -5$

$4x - 6y = 2$

❽

$x - 7y = 4$

$-2x + y = 9$

❾

$4x - 9y = -9$

$x + 2y = 4$

❿

$7x + 4y = 4$

$-x + 7y = -2$

⓫

$-4x - 5y = -3$

$6x + 2y = -2$

⓬

$3x - 3y = 5$

$x + 2y = -1$

❶

$-4x - 3y = 5$

$-5x - 7y = 5$

❷

$3x + 2y = 8$

$2x + 5y = -4$

❸

$x - 3y = 8$

$-6x - y = 9$

❹

$-8x + 5y = -8$

$8x - 6y = 5$

❺

$5x - 6y = 4$

$5x - 4y = 2$

❻

$-4x - 6y = -7$

$-7x - 9y = -3$

❼

$-2x - 2y = 7$

$-4x - 8y = -4$

❽

$2x + 7y = 9$

$5x + 6y = -6$

❾

$-x + 2y = 2$

$5x - 4y = 1$

❿

$-9x + y = -8$

$x - y = -7$

⓫

$-5x - 7y = -6$

$-x + 7y = -2$

⓬

$5x - 6y = 9$

$-6x - 8y = -3$

❶

$-7x - 2y = -7$

$7x - 4y = -2$

❷

$x - 2y = -7$

$-7x - 7y = -8$

❸

$-x - 3y = 7$

$-9x - 6y = 8$

❹

$-x + 5y = 5$

$2x - 3y = 8$

❺

$7x - y = 6$

$6x + 2y = 9$

❻

$5x + 7y = 8$

$6x + 4y = -7$

❼

$-9x + 6y = -7$

$4x + 8y = -3$

❽

$-8x - y = -2$

$-8x + 7y = 9$

❾

$-9x + 6y = 6$

$-9x + 4y = -9$

❿

$-x - 9y = -8$

$-7x - 9y = 2$

⓫

$4x + 3y = 3$

$x - 2y = 4$

⓬

$-5x + 2y = 5$

$-5x + 9y = 7$

❶

$-5x + 4y = 1$

$-7x + 7y = -7$

❷

$-8x - y = -3$

$8x - 3y = 2$

❸

$2x - 5y = -7$

$5x + 2y = 1$

❹

$-x - 6y = 3$

$9x + 9y = -2$

❺

$-2x + 6y = 8$

$x + 7y = -6$

❻

$-2x + 8y = 2$

$x + 3y = 4$

❼

$-x + 2y = -8$

$7x + y = 7$

❽

$2x - 3y = 9$

$8x - 7y = -5$

❾

$-6x - 3y = 2$

$x + 6y = -2$

❿

$5x + y = 3$

$4x + 6y = 4$

⓫

$2x + 3y = -6$

$3x + 4y = 3$

⓬

$x + y = -3$

$-x + 5y = -1$

❶
$-6x - 4y = 7$
$-3x + 9y = 2$

❷
$2x - 8y = -5$
$5x - 5y = -6$

❸
$-x + 5y = 8$
$-4x + 8y = -6$

❹
$-8x - 8y = 9$
$-7x - 4y = -4$

❺
$x - 4y = -9$
$9x - 8y = -4$

❻
$-6x - 6y = 8$
$-4x + 6y = 2$

❼
$9x - 9y = -7$
$-5x + y = -9$

❽
$7x + 6y = -3$
$-x - y = 5$

❾
$-8x + y = -3$
$8x - 9y = -2$

❿
$-3x + 3y = 7$
$-4x - 5y = 4$

⓫
$9x - 5y = 2$
$2x + y = -3$

⓬
$-5x + 6y = -6$
$7x - 5y = -3$

❶

$7x - 5y = 3$

$5x + 9y = 2$

❷

$7x + 7y = -1$

$-9x - 2y = 6$

❸

$5x + y = -3$

$-x + 5y = 2$

❹

$-3x + 2y = 6$

$4x + 8y = -7$

❺

$9x - 3y = -5$

$-7x + 8y = -6$

❻

$x + y = -7$

$5x + 8y = -9$

❼

$8x - 5y = 7$

$-9x + y = -3$

❽

$6x - 7y = 4$

$-7x + y = -2$

❾

$-3x - y = 5$

$8x - 5y = -5$

❿

$5x - y = -8$

$-8x + 6y = -7$

⓫

$-4x - 2y = 1$

$-5x - 7y = 8$

⓬

$6x - 7y = 1$

$7x - 6y = 6$

❶

$-6x - 7y = -3$

$-x - 6y = -7$

❷

$6x - 2y = -7$

$-8x + 6y = 8$

❸

$-7x - 2y = 8$

$2x + 4y = 7$

❹

$5x + y = -8$

$9x - y = -1$

❺

$-3x - 2y = -6$

$x + 3y = -5$

❻

$-3x - 3y = -7$

$-4x - 6y = -8$

❼

$-4x + 7y = -4$

$-5x - 7y = 8$

❽

$-4x - 9y = -5$

$-7x + y = 2$

❾

$3x - y = -6$

$-6x + 7y = 9$

❿

$-3x + 8y = 7$

$6x + 6y = 7$

⓫

$-7x - 3y = -9$

$7x - 9y = 6$

⓬

$-7x - 4y = 6$

$-6x + 3y = -9$

❶

$-x + 7y = 9$

$-x + 6y = 8$

❷

$4x - 7y = -6$

$-x + 9y = 9$

❸

$-9x + y = -3$

$8x + 4y = -6$

❹

$3x - y = 6$

$-7x + 5y = 1$

❺

$7x + 8y = 5$

$-7x + 6y = -4$

❻

$-6x - 2y = 2$

$-6x - 9y = -8$

❼

$6x - y = -4$

$7x + 6y = -4$

❽

$7x + y = 3$

$-7x + 2y = 6$

❾

$8x - 2y = 6$

$7x - 5y = -6$

❿

$3x - 7y = -4$

$-4x + 9y = 7$

⓫

$2x + y = 5$

$x - y = -8$

⓬

$3x + 2y = 1$

$5x - 8y = 5$

❶

$x - 7y = -2$

$-4x - 8y = 8$

❷

$-2x - 8y = 1$

$2x + 2y = 8$

❸

$-3x - 6y = 4$

$5x + 4y = 3$

❹

$-7x - 8y = -3$

$2x + y = 2$

❺

$5x + 5y = 8$

$-9x + 3y = 6$

❻

$-4x + 4y = -9$

$7x - 3y = -8$

❼

$-2x - 2y = -4$

$-x + 7y = -4$

❽

$8x - 7y = -7$

$x - 2y = -4$

❾

$7x + 3y = 4$

$4x + 4y = -7$

❿

$2x + 2y = -3$

$-6x + 4y = -8$

⓫

$9x + 5y = 3$

$-3x + 4y = -1$

⓬

$7x + 3y = -4$

$x - 5y = 9$

❶

$9x + 4y = 9$

$-4x + y = -4$

❷

$5x - 8y = -4$

$9x - y = 2$

❸

$5x - y = 1$

$5x + 7y = -3$

❹

$-x + 6y = -4$

$x + 4y = -6$

❺

$-3x - 9y = 7$

$8x + 9y = -1$

❻

$7x - 3y = 9$

$8x - 3y = 4$

❼

$-4x - 3y = 8$

$-8x + 8y = -8$

❽

$x + 5y = 5$

$-5x - 5y = -6$

❾

$-4x + 4y = -5$

$3x + y = 5$

❿

$x + 3y = -6$

$-5x - 2y = -6$

⓫

$5x + y = 9$

$4x - 4y = -1$

⓬

$2x - 6y = -4$

$-6x - 4y = 8$

Answer Key

Chapter 1 Answers:

Page 8
❶ −17/2 ❷ −1 ❸ −2/7
❹ −2 ❺ 1/14 ❻ 2
❼ −1/5 ❽ 3 ❾ −2/3
❿ 8/5 ⓫ −1 ⓬ −3/13
⓭ −5 ⓮ −6/7 ⓯ 1/2
⓰ 7/12 ⓱ −4 ⓲ 1

Page 9
❶ −3 ❷ 9/16 ❸ 1/3
❹ 5 ❺ −5/16 ❻ −9
❼ −3 ❽ −3/5 ❾ −2
❿ −7/6 ⓫ −4/13 ⓬ 0
⓭ 8/7 ⓮ 8/5 ⓯ −5/3
⓰ 5/4 ⓱ 1/3 ⓲ 0

Page 10
❶ −9/4 ❷ 1/2 ❸ −1/5
❹ 13/2 ❺ 2/3 ❻ 6/11
❼ 1 ❽ 5/7 ❾ −3/5
❿ 3/8 ⓫ −3/8 ⓬ 2
⓭ 3/4 ⓮ −4 ⓯ −1/9
⓰ 4/5 ⓱ 1/8 ⓲ 4/5

Page 11
❶ −3/10 ❷ −12 ❸ 1/7
❹ 7/9 ❺ −8/3 ❻ 4/5
❼ −1 ❽ −9/10 ❾ 3/17
❿ −4/13 ⓫ 9/8 ⓬ −1/2
⓭ −1 ⓮ 3/13 ⓯ 9/14
⓰ 13/3 ⓱ 9/11 ⓲ −4

Page 12
❶ −8/3 ❷ 1 ❸ 1/2
❹ −3 ❺ 2 ❻ 0
❼ 6/11 ❽ 7/4 ❾ −1/2
❿ −13/6 ⓫ 1/2 ⓬ −3/4
⓭ 11 ⓮ −1 ⓯ −2
⓰ −1/3 ⓱ 1/5 ⓲ −2/3

Page 13
❶ −1/10 ❷ −1/5 ❸ 7/10
❹ −5/3 ❺ −2 ❻ 1/8
❼ 11/2 ❽ 2/5 ❾ 0
❿ −7/5 ⓫ 2/3 ⓬ −5
⓭ 0 ⓮ 11/6 ⓯ 3/4
⓰ −4/13 ⓱ 0 ⓲ −5/3

Page 14
❶ 9/8 ❷ −13 ❸ −8/13
❹ 0 ❺ −2 ❻ −1/18
❼ 1/2 ❽ 8 ❾ 6/11
❿ 5/6 ⓫ −2/7 ⓬ −2/11
⓭ −13/3 ⓮ −2/3 ⓯ 4/3
⓰ 3/2 ⓱ −7/6 ⓲ 1/3

Page 15
❶ −1/26 ❷ −9 ❸ −1/3
❹ 6/7 ❺ 0 ❻ 5/14
❼ 0 ❽ −5/7 ❾ −1
❿ 5 ⓫ 1/14 ⓬ 4/5
⓭ 7/13 ⓮ −3/4 ⓯ 2/5
⓰ 8/5 ⓱ −5/6 ⓲ 1/5

Page 16
❶ −11/9 ❷ −7/4 ❸ 2/3
❹ −1 ❺ 9/13 ❻ −4/5
❼ −1/7 ❽ 5/8 ❾ 0
❿ 9/14 ⓫ −3/4 ⓬ 1
⓭ 0 ⓮ 9 ⓯ −13/6
⓰ 3/2 ⓱ 4/3 ⓲ −8/11

Page 17
❶ −2 ❷ −5/2 ❸ −9/2
❹ −1 ❺ −1/5 ❻ 1
❼ 7/9 ❽ −8/5 ❾ 1/3
❿ 6/7 ⓫ −12/5 ⓬ 1
⓭ 1 ⓮ −4/3 ⓯ −1
⓰ 0 ⓱ −1/10 ⓲ 1/6

Page 18
❶ 7/6 ❷ −1 ❸ 5
❹ 0 ❺ −9/11 ❻ 9/4
❼ 2 ❽ 5 ❾ −1
❿ 5/2 ⓫ −6 ⓬ 6/5
⓭ 4/11 ⓮ 0 ⓯ 1/7
⓰ −1/4 ⓱ −5/17 ⓲ −17/8

Page 19
❶ −3/5 ❷ 1/4 ❸ −1/2
❹ −9/2 ❺ 2/7 ❻ 1
❼ 1/11 ❽ 9/4 ❾ 1/8
❿ −8/5 ⓫ 7/18 ⓬ 5/14
⓭ 7/3 ⓮ 1/4 ⓯ −3
⓰ 13/6 ⓱ 1/3 ⓲ −3/2

Page 20
❶ −3/5 ❷ 3/4 ❸ 1
❹ −14/15 ❺ 2 ❻ 0
❼ −2/5 ❽ −1 ❾ −3
❿ 5/13 ⓫ −5/8 ⓬ 0
⓭ −13/3 ⓮ 8/3 ⓯ −8/5
⓰ −6 ⓱ 7/11 ⓲ −8/11

Page 21
❶ 11/15 ❷ −2 ❸ 2/7
❹ 6/7 ❺ 2/7 ❻ −3
❼ −5/16 ❽ 7/2 ❾ 8/5
❿ −1 ⓫ 3/2 ⓬ −5/9
⓭ 1 ⓮ −9/2 ⓯ 15/2
⓰ −1 ⓱ 14 ⓲ −3/16

Page 22
❶ 0 ❷ 3/16 ❸ −2
❹ −9/4 ❺ −1/2 ❻ 6
❼ 5/4 ❽ −9/2 ❾ −1/2
❿ 1/8 ⓫ 2 ⓬ −9/4
⓭ −3/4 ⓮ 2/5 ⓯ 1/3
⓰ 1/8 ⓱ −2 ⓲ −1

Page 23
❶ 3/16 ❷ 0 ❸ 0
❹ 3/13 ❺ 7/5 ❻ 8/5
❼ 2/11 ❽ −6 ❾ −4/13
❿ −3/4 ⓫ 1 ⓬ −3/2
⓭ −4/11 ⓮ 2 ⓯ 7/9
⓰ −17/11 ⓱ 0 ⓲ −7/5

Page 24
❶ 2/5 ❷ 9/11 ❸ 3
❹ 3/2 ❺ −1/2 ❻ 11/10
❼ −3/4 ❽ −10/3 ❾ −4
❿ −8/13 ⓫ −4/3 ⓬ −1/4
⓭ 6/7 ⓮ −5/4 ⓯ −1/3
⓰ 1/2 ⓱ 3/7 ⓲ 0

Page 25
❶ −2/3 ❷ −4/7 ❸ −5/2
❹ −5/11 ❺ 3/7 ❻ −11/12
❼ −1/3 ❽ −1/8 ❾ 11/6
❿ −4 ⓫ −1/3 ⓬ 2/3
⓭ −4/17 ⓮ 3/7 ⓯ 5
⓰ −9/2 ⓱ −5/18 ⓲ −5/2

Page 26
❶ 2/3 ❷ −7 ❸ −4/3
❹ 14 ❺ 3 ❻ −7/2
❼ −1 ❽ 9 ❾ −9/8
❿ 1 ⓫ 3/2 ⓬ −8/3
⓭ 7/13 ⓮ −1/2 ⓯ 0
⓰ −8/9 ⓱ 6/11 ⓲ −16/3

Page 27
❶ −8/9 ❷ −1/5 ❸ 0
❹ −9/11 ❺ 7 ❻ 7/9
❼ −3/5 ❽ −2/7 ❾ 11/3
❿ 1/11 ⓫ 5/12 ⓬ 5/7
⓭ −2 ⓮ −5/4 ⓯ 1/11
⓰ 2 ⓱ 1 ⓲ −2/5

Chapter 2 Answers:

Page 31
❶ 5/6 ❷ 2/9 ❸ 40/3
❹ –76/55 ❺ 2/7 ❻ 16/45
❼ –1/11 ❽ 9/16 ❾ 9/16
❿ –9/29 ⓫ –9/20 ⓬ –31/30
⓭ 16/3 ⓮ –30/23 ⓯ 25/21
⓰ 2/3 ⓱ 1/6 ⓲ 5/16

Page 32
❶ 11/12 ❷ 1/8 ❸ –87/20
❹ 2/5 ❺ –10/9 ❻ –3/5
❼ –5/6 ❽ 2 ❾ 11/14
❿ 3/2 ⓫ 0 ⓬ –20/13
⓭ –4/3 ⓮ 5/2 ⓯ 0
⓰ 95/13 ⓱ –1/11 ⓲ 15/16

Page 33
❶ 9/4 ❷ 0 ❸ –4
❹ –6/5 ❺ 4/3 ❻ 0
❼ –4/5 ❽ –45/19 ❾ 5/68
❿ –1/20 ⓫ –1/2 ⓬ –6/5
⓭ 5/6 ⓮ 25/99 ⓯ –10/17
⓰ 2/3 ⓱ 26/15 ⓲ 6/5

Page 34
❶ 13/5 ❷ –10 ❸ –3/2
❹ –5/4 ❺ 4/15 ❻ –32/25
❼ 0 ❽ –3/2 ❾ –8/21
❿ –2/5 ⓫ 16/55 ⓬ –11/8
⓭ 0 ⓮ 19/10 ⓯ 40/3
⓰ 8/5 ⓱ 1/3 ⓲ 25/24

Page 35
❶ 0 ❷ –8/41 ❸ 1/3
❹ 22/9 ❺ 5/44 ❻ 3/4
❼ 0 ❽ 6/13 ❾ –5/18
❿ 5/3 ⓫ –1/8 ⓬ –45/11
⓭ 8/43 ⓮ 4/11 ⓯ 0
⓰ –20/3 ⓱ –5/3 ⓲ 3/11

Page 36
❶ 84/85 ❷ –3 ❸ 15/11
❹ –1/5 ❺ –20/13 ❻ 8/11
❼ –9/5 ❽ 12/5 ❾ –15/2
❿ –2/5 ⓫ 3/32 ⓬ –5/4
⓭ –2 ⓮ –8/3 ⓯ 6/55
⓰ –4/5 ⓱ –85/4 ⓲ 15/4

Page 37
❶ 1/15 ❷ –20/17 ❸ –15/14
❹ –2/17 ❺ 0 ❻ –15/46
❼ –9/35 ❽ 0 ❾ 3/2
❿ –5/2 ⓫ 4/5 ⓬ –3/5
⓭ 3/2 ⓮ 9/10 ⓯ –1/5
⓰ –16/3 ⓱ 0 ⓲ –3/7

Page 38
❶ 15/73 ❷ 2/3 ❸ 9/5
❹ –4 ❺ 0 ❻ 15/26
❼ 4/9 ❽ –35/33 ❾ 6
❿ 55/58 ⓫ –5/18 ⓬ 15/2
⓭ 5 ⓮ –3/2 ⓯ 60
⓰ 2 ⓱ –50/13 ⓲ –1/16

Page 39
❶ –50 ❷ 4 ❸ 4/5
❹ –12/5 ❺ –3/10 ❻ –75
❼ –9/23 ❽ –69/10 ❾ 45/2
❿ –4 ⓫ –7/3 ⓬ –9/14
⓭ 8/35 ⓮ 78/25 ⓯ –3/2
⓰ 4/23 ⓱ 3 ⓲ –1

Page 40
❶ –45/41 ❷ –1/5 ❸ 1/35
❹ 9/14 ❺ 1/9 ❻ –3/8
❼ 13/16 ❽ 0 ❾ 18/11
❿ 10/17 ⓫ –19/15 ⓬ –5/26
⓭ 22/39 ⓮ 16/51 ⓯ 2/7
⓰ 4/3 ⓱ 1/4 ⓲ 1/5

Page 41
❶ −3/13 ❷ −19/12 ❸ −35/36
❹ −9/5 ❺ −21/22 ❻ 9/10
❼ −10/13 ❽ 0 ❾ 2/3
❿ 4/7 ⓫ −5/6 ⓬ 18/19
⓭ 4/31 ⓮ −1/2 ⓯ 1/5
⓰ 1 ⓱ 18/41 ⓲ 12/11

Page 42
❶ −12 ❷ 25/16 ❸ −4/35
❹ 2/5 ❺ −6/19 ❻ −5/3
❼ 20/9 ❽ −21/50 ❾ 9/20
❿ −24 ⓫ 2/9 ⓬ 25/14
⓭ 14/3 ⓮ 5/4 ⓯ −2/9
⓰ −25/26 ⓱ 23/24 ⓲ 10/3

Page 43
❶ 2 ❷ 0 ❸ −8/5
❹ −10 ❺ −36/5 ❻ 21/25
❼ 45/88 ❽ 0 ❾ −3
❿ −5/57 ⓫ −12/5 ⓬ −36/31
⓭ −7/4 ⓮ 2/7 ⓯ −4/13
⓰ 26/7 ⓱ −33/20 ⓲ −65/24

Page 44
❶ −4/5 ❷ 2 ❸ 3
❹ −18/65 ❺ 0 ❻ −6
❼ 8/11 ❽ −6/35 ❾ −1
❿ 3/11 ⓫ −17/8 ⓬ −1/3
⓭ 1/9 ⓮ −30/11 ⓯ −2
⓰ 3/2 ⓱ −1/90 ⓲ −17/20

Page 45
❶ 1/3 ❷ 15/4 ❸ −1/20
❹ −1/3 ❺ 5/14 ❻ −56/15
❼ 5/8 ❽ 85/3 ❾ −2
❿ 3 ⓫ 1/9 ⓬ −8/3
⓭ 6/13 ⓮ 60/71 ⓯ −20/83
⓰ −1 ⓱ −5/36 ⓲ −12/5

Page 46
❶ −3/7 ❷ −11/2 ❸ −2
❹ −5/2 ❺ −3/5 ❻ −14/17
❼ 0 ❽ 1/3 ❾ 0
❿ 1/4 ⓫ −6/13 ⓬ −5/24
⓭ 48/41 ⓮ −15/17 ⓯ 22/45
⓰ −1 ⓱ 75/22 ⓲ −24

Page 47
❶ 2/7 ❷ 16/27 ❸ −18/41
❹ 3/8 ❺ 35/6 ❻ 18/31
❼ −1/12 ❽ 75/92 ❾ −23/8
❿ −9/10 ⓫ 3/5 ⓬ −1/6
⓭ 15/53 ⓮ 12/25 ⓯ 12/17
⓰ 7/45 ⓱ 12/59 ⓲ −3/20

Page 48
❶ −36/5 ❷ 21/40 ❸ −1/4
❹ 54/7 ❺ 6/5 ❻ 4/11
❼ −22 ❽ −15/2 ❾ −24/7
❿ 3/38 ⓫ −1 ⓬ −1/5
⓭ −1 ⓮ −2/39 ⓯ −3/7
⓰ 10/93 ⓱ −61/50 ⓲ 5/4

Page 49
❶ 9/10 ❷ 5/9 ❸ 5/58
❹ 3/10 ❺ −2/5 ❻ 12
❼ −1/2 ❽ −5/22 ❾ 9/4
❿ −12 ⓫ 4/3 ⓬ 0
⓭ −6/25 ⓮ 0 ⓯ −2
⓰ −2/7 ⓱ 3/10 ⓲ 28/25

Page 50
❶ 1 ❷ −10/3 ❸ −12
❹ −5/24 ❺ −19/13 ❻ −4
❼ 20/9 ❽ −60/23 ❾ 22/39
❿ 7/4 ⓫ 32/5 ⓬ 3/32
⓭ −25/8 ⓮ 12/61 ⓯ −1
⓰ −5/2 ⓱ −27/20 ⓲ −55/6

Chapter 3 Answers:

Page 56

❶ 0 ❷ $\pm\sqrt{6}$ ❸ 0, –7/22
❹ ±1 ❺ 0, 14 ❻ 0, 2
❼ 0, 8 ❽ 0, 7 ❾ $\pm\sqrt{35}/5$
❿ 0 ⓫ 0, –3/2 ⓬ 0, –2/3
⓭ $\pm\sqrt{42}/3$ ⓮ 0, –4/15 ⓯ 0, 1/2
⓰ 0, 1/18 ⓱ 0, 9/14 ⓲ $\pm\sqrt{3}$

Page 57

❶ 0, 7/4 ❷ $\pm\sqrt{22}/11$ ❸ $\pm\sqrt{5}/5$
❹ 0, 8 ❺ ±1/4 ❻ 0
❼ $\pm\sqrt{15}/5$ ❽ 0, 6/11 ❾ 0, 1/10
❿ 0, 4/13 ⓫ 0, 11/6 ⓬ 0, –1
⓭ 0, 1 ⓮ $\pm\sqrt{7}$ ⓯ 0
⓰ 0, –7 ⓱ 0, 2/3 ⓲ 0, 4

Page 58

❶ 0, –7 ❷ 0 ❸ $\pm\sqrt{6}/2$
❹ 0, –3/5 ❺ 0 ❻ $\pm\sqrt{7}$
❼ 0, 6 ❽ 0, –7/11 ❾ 0, –3/7
❿ 0 ⓫ 0, 6/7 ⓬ $\pm\sqrt{3}/3$
⓭ 0, 1 ⓮ 0, –2/3 ⓯ $\pm\sqrt{2}$
⓰ ±1 ⓱ 0 ⓲ 0, –1

Page 59

❶ 0, 5/2 ❷ $\pm\sqrt{10}/5$ ❸ 0, –2/3
❹ 0, 2/5 ❺ 0, –1/14 ❻ 0, –5
❼ 0 ❽ 0, 7/15 ❾ 0
❿ 0, –1/15 ⓫ 0, –1 ⓬ 0, 4/9
⓭ 0, 13/15 ⓮ $\pm\sqrt{5}$ ⓯ 0
⓰ 0, –13/12 ⓱ 0 ⓲ 0, 10/7

Page 60

❶ $\pm\sqrt{14}/2$ ❷ 0, –2/17 ❸ 0, –2/3
❹ 0, –1/8 ❺ 0, –3/11 ❻ $\pm\sqrt{7}/3$
❼ 0, –5/13 ❽ 0, –1 ❾ 0
❿ 0, –2/5 ⓫ 0, –8/9 ⓬ $\pm\sqrt{2}/3$
⓭ $\pm\sqrt{3}$ ⓮ 0 ⓯ 0, –2/7
⓰ $\pm\sqrt{6}$ ⓱ 0, –11/4 ⓲ $\pm 2\sqrt{39}/13$

Page 61

❶ 0 ❷ $\pm 2\sqrt{7}/7$ ❸ $\pm\sqrt{102}/17$
❹ $\pm\sqrt{6}/3$ ❺ 0, 2/3 ❻ 0, –3/2
❼ 0, 15/2 ❽ ±3/2 ❾ $\pm\sqrt{3}$
❿ 0, –1/2 ⓫ 0, –9/16 ⓬ $\pm\sqrt{5}/4$
⓭ 0, 4/3 ⓮ $\pm 2\sqrt{5}/5$ ⓯ 0, –2/9
⓰ 0, 2/3 ⓱ 0, –9/8 ⓲ $\pm\sqrt{2}$

Page 62

❶ 0 ❷ 0, 8/3 ❸ ±1/2
❹ $\pm\sqrt{35}/5$ ❺ 0, 8/11 ❻ 0, –5/3
❼ 0, 1/3 ❽ ±1 ❾ 0, –3/4
❿ $\pm\sqrt{15}/6$ ⓫ 0, 2 ⓬ 0, –1/15
⓭ 0, 3/2 ⓮ 0 ⓯ 0
⓰ 0 ⓱ 0, –1/8 ⓲ 0, 8/9

Page 63

❶ 0, 3/13 ❷ 0, –8/5 ❸ 0, 4/3
❹ 0, –5/7 ❺ 0, –1/7 ❻ 0, 2
❼ 0, –1 ❽ 0, –2 ❾ 0
❿ 0, –1/6 ⓫ 0, –5/12 ⓬ 0, –1/4
⓭ 0, 9/10 ⓮ 0, –3/2 ⓯ 0, –1/10
⓰ ±1/3 ⓱ $\pm\sqrt{2}/3$ ⓲ 0, –6/13

Page 64

❶ 0 ❷ 0 ❸ 0
❹ 0, 1/3 ❺ $\pm 2\sqrt{42}/21$ ❻ 0, –1/16
❼ 0, –4/9 ❽ 0, –1 ❾ 0, 5/2
❿ 0, 1/20 ⓫ 0, 3/2 ⓬ 0, 1
⓭ 0, 8/3 ⓮ 0 ⓯ 0, 4/7
⓰ $\pm 3\sqrt{2}/2$ ⓱ 0, 1/6 ⓲ ±1

Page 65

❶ $\pm\sqrt{10}/2$ ❷ 0, 3/8 ❸ 0, –1
❹ $\pm\sqrt{7}$ ❺ 0, 4 ❻ $\pm\sqrt{35}/7$
❼ 0, 7/5 ❽ 0, 2/5 ❾ ±2
❿ 0, –3/2 ⓫ 0, 1/11 ⓬ 0, 1/2
⓭ $\pm 2\sqrt{5}/5$ ⓮ $\pm\sqrt{13}$ ⓯ $\pm\sqrt{10}/2$
⓰ 0, –5/7 ⓱ 0, –2 ⓲ 0, –9/7

Page 66

❶ 0, −2/5 ❷ ±1 ❸ 0, −5/9
❹ $\pm\sqrt{5}$ ❺ $\pm\sqrt{10}/4$ ❻ 0, −2/3
❼ $\pm\sqrt{11}/2$ ❽ 0, −7/11 ❾ $\pm\sqrt{3}/2$
❿ 0, 9/5 ⓫ 0, −9 ⓬ 0, 16
⓭ 0, −4/5 ⓮ $\pm\sqrt{3}/2$ ⓯ $\pm\sqrt{6}/6$
⓰ 0, −1/13 ⓱ 0, 8/5 ⓲ ±1/2

Page 67

❶ 0, 8/13 ❷ $\pm\sqrt{15}/5$ ❸ 0
❹ 0, 2/5 ❺ 0, −5/9 ❻ 0, −2/3
❼ $\pm 2\sqrt{2}$ ❽ 0, 2/3 ❾ 0, −4
❿ 0, 3 ⓫ $\pm\sqrt{7}/2$ ⓬ $\pm\sqrt{21}/3$
⓭ 0, 7/13 ⓮ 0, 2/3 ⓯ ±2
⓰ 0, 1/2 ⓱ 0, 2 ⓲ 0

Page 68

❶ 0, 8/5 ❷ 0, 1/3 ❸ $\pm\sqrt{3}$
❹ 0 ❺ 0, −3/11 ❻ 0, −9/4
❼ 0, 4 ❽ 0, 1 ❾ 0, −9/4
❿ 0, 6/5 ⓫ 0, −1/6 ⓬ 0, 4/11
⓭ $\pm\sqrt{3}/3$ ⓮ 0, 1/6 ⓯ 0, 6
⓰ $\pm\sqrt{3}/3$ ⓱ 0, 4/3 ⓲ 0, 13/5

Page 69

❶ 0, 5/2 ❷ 0, 4 ❸ 0
❹ $\pm\sqrt{15}/6$ ❺ $\pm\sqrt{30}/6$ ❻ ±1
❼ $\pm\sqrt{30}/5$ ❽ 0 ❾ 0, −5/3
❿ 0, 3/7 ⓫ 0, 3 ⓬ $\pm\sqrt{6}/4$
⓭ 0, 5/9 ⓮ 0 ⓯ 0, 13/12
⓰ 0, −1/4 ⓱ 0, −8/15 ⓲ $\pm 2\sqrt{5}/5$

Page 70

❶ 0, 7 ❷ 0, −3/5 ❸ 0, −1/4
❹ 0, 1/11 ❺ 0, −3/11 ❻ 0, −9/8
❼ 0, 1 ❽ $\pm\sqrt{6}/3$ ❾ 0, −16/3
❿ 0, 7 ⓫ 0 ⓬ 0, 2/5
⓭ 0, 7 ⓮ 0, 2/3 ⓯ 0
⓰ 0, −1/3 ⓱ 0, −3/2 ⓲ 0

Page 71

❶ 0, −1/18 ❷ $\pm\sqrt{3}/3$ ❸ 0, 16/5
❹ $\pm 2\sqrt{3}/3$ ❺ $\pm 2\sqrt{5}/5$ ❻ 0, −17/11
❼ 0 ❽ 0, −9/11 ❾ $\pm\sqrt{10}$
❿ 0, 3/2 ⓫ 0 ⓬ 0, −7/4
⓭ 0, −1 ⓮ 0, 7/2 ⓯ 0, −9/13
⓰ $\pm 2\sqrt{39}/13$ ⓱ 0, −16/9 ⓲ 0, 6

Page 72

❶ $\pm\sqrt{10}/2$ ❷ 0, −4/15 ❸ 0, −5/7
❹ $\pm 2\sqrt{26}/13$ ❺ 0, −13/14 ❻ 0, 6
❼ 0, −1/4 ❽ $\pm\sqrt{21}/3$ ❾ 0, −4/3
❿ 0, −5/2 ⓫ $\pm\sqrt{42}/6$ ⓬ 0, 2/3
⓭ 0, 3/5 ⓮ $\pm\sqrt{6}/6$ ⓯ ±1
⓰ 0, 7/5 ⓱ 0, −4/9 ⓲ 0, −1/8

Page 73

❶ 0, −1/7 ❷ 0, 3 ❸ 0, 2
❹ $\pm 2\sqrt{10}/5$ ❺ $\pm\sqrt{35}/5$ ❻ 0, −17/12
❼ 0, −1/3 ❽ 0, −1/2 ❾ ±1
❿ 0, 3/2 ⓫ 0, 1/2 ⓬ $\pm\sqrt{30}/6$
⓭ $\pm\sqrt{5}/4$ ⓮ ±1/5 ⓯ $\pm\sqrt{21}/6$
⓰ 0 ⓱ $\pm 3\sqrt{11}/11$ ⓲ $\pm 3\sqrt{13}/13$

Page 74

❶ 0, 1/2 ❷ 0, −2 ❸ 0, −2
❹ 0, −1 ❺ 0, −2/9 ❻ 0, 4/3
❼ $\pm 2\sqrt{10}/5$ ❽ ±1 ❾ 0, 3/5
❿ 0, 8 ⓫ ±1 ⓬ 0, −3/2
⓭ $\pm\sqrt{35}/5$ ⓮ 0, −3/8 ⓯ $\pm\sqrt{3}/3$
⓰ 0, 5/4 ⓱ 0, −3 ⓲ 0, 7/6

Page 75

❶ 0 ❷ $\pm\sqrt{6}/2$ ❸ ±2
❹ $\pm\sqrt{7}/7$ ❺ $\pm\sqrt{7}$ ❻ 0, −1/9
❼ 0, 1/2 ❽ $\pm\sqrt{5}/2$ ❾ $\pm 3\sqrt{11}/11$
❿ $\pm\sqrt{10}/2$ ⓫ 0 ⓬ 0, −2
⓭ 0, −2/3 ⓮ 0, −1/2 ⓯ 0, 1/4
⓰ $\pm\sqrt{2}/2$ ⓱ 0, 8/3 ⓲ 0, 9/2

Chapter 4 Answers:

Page 79

❶ 1/2, 1 ❷ −3, 2/5 ❸ −4, 6
❹ −7/5, 5 ❺ −3/4, −5/2 ❻ 6, 2
❼ −1/2, −2 ❽ −9/2, −7/3 ❾ −1, 5/4
❿ −2, 6/5 ⓫ 9, 3/2 ⓬ −3, 7/4

Page 80

❶ 8, −2 ❷ 3, 6/5 ❸ −8/5, −1/2
❹ 7/4, −3/2 ❺ 5/3, 1 ❻ −1, −7/2
❼ −4/3, 1 ❽ 3/5, −3 ❾ −3/4, 3
❿ −1/5, 6 ⓫ −5/4, −7/2 ⓬ −6/5, 8/3

Page 81

❶ 8, −1/3 ❷ −4, −2/5 ❸ −7/2, −6
❹ −8/5, 7/4 ❺ 7/3, −4 ❻ 9/4, −8/5
❼ −2/3, 1/2 ❽ −7/5, 4 ❾ 3/5, 9/4
❿ −2/3, −2 ⓫ 1, 7/2 ⓬ 9/4, 4

Page 82

❶ −2, 8/3 ❷ −2, 1/2 ❸ −8/5, −5/4
❹ −7/5, 1/2 ❺ −1, 4/3 ❻ 1/3, 2/5
❼ −7/4, 3 ❽ 2, 3/2 ❾ −1, 1
❿ 3/2, −1 ⓫ −2, −1 ⓬ −1/4, −6

Page 83

❶ −7/2, −3/2 ❷ −7, −7 ❸ −3/5, 7/5
❹ −5/4, −3 ❺ −1, 1 ❻ 1, 8/5
❼ −6, −5/4 ❽ −5/3, −8/5 ❾ 3, −1/5
❿ −3, 1 ⓫ 4, 8/3 ⓬ −2/3, −8

Page 84

❶ −3/2, −4/5 ❷ 7/4, −6/5 ❸ −1, −5/4
❹ −3, −7/5 ❺ −1/4, −3/2 ❻ 1/5, −1
❼ 5, 4/5 ❽ 2, 2 ❾ 1, −5
❿ 3, −2 ⓫ 1, 2/5 ⓬ −2, 8/5

Page 85

❶ −1, 6 ❷ 3/2, −8/5 ❸ 2, −9/5
❹ −1, 1 ❺ 6, −4/5 ❻ −9/2, 1
❼ 5/2, −2 ❽ 6, −1 ❾ −7, 1/5
❿ −8/3, 8/3 ⓫ 2, 1 ⓬ −6, −1

Page 86

❶ −1/3, 3/2 ❷ 9/2, −4/3 ❸ 4, −2/3
❹ 1, 1 ❺ 3/2, −1/2 ❻ 9/5, 5/2
❼ −2, −1/2 ❽ 2, 4 ❾ 5/3, −1/2
❿ −1, −7/3 ⓫ 4, 1 ⓬ −1/2, −7/3

Page 87

❶ 1, −2/3 ❷ −5, −1/5 ❸ 2, 4/5
❹ −7/4, −1 ❺ −8/3, 1/3 ❻ −5, 5/4
❼ −3, −5/4 ❽ −3, −1/3 ❾ −1/4, −3/2
❿ −1/5, 1/5 ⓫ 1/2, −4/3 ⓬ −8/5, 1

Page 88

❶ 3/4, 5 ❷ 7/2, 3 ❸ 4, 6/5
❹ 2, 8/5 ❺ −9/5, −1/2 ❻ −8, −1/4
❼ −3/2, 8/5 ❽ −5/4, 7/3 ❾ 5/4, −3
❿ 3/5, −4 ⓫ 9/2, −1/2 ⓬ 3, −1/2

Page 89

❶ 4, −6 ❷ 3, 3/5 ❸ 1, −1/2
❹ 2/5, 1/2 ❺ 2, −6 ❻ 2, 9/4
❼ −7/2, 2 ❽ −2, 1 ❾ 3, 3
❿ 9/2, −2 ⓫ −1, 7/3 ⓬ −6/5, −3/2

Page 90

❶ −1, 2 ❷ 7, 2/3 ❸ 7/5, −1/5
❹ −1/5, −9 ❺ −1/5, 1 ❻ −1, −1/3
❼ 1/3, 7/4 ❽ −3, −7/3 ❾ 7/5, −5/4
❿ −9/4, −1 ⓫ 9/2, 8/5 ⓬ 3, 7/2

Page 91

❶ 1, 1/4 ❷ 9/5, −1/4 ❸ −7, 9
❹ −4, 1 ❺ −1/2, 2 ❻ 4, 2
❼ 6, −4/5 ❽ −1, −9/5 ❾ 9, −9/2
❿ −8/3, −8/5 ⓫ 3/5, −4 ⓬ 2/3, −8

Page 92

❶ 8/3, −1/2 ❷ 9/2, −1/4 ❸ 4, −9/2
❹ −2, 4/3 ❺ −2, −1 ❻ −2/3, −3
❼ 2, −2 ❽ −2/3, −7/3 ❾ −1/2, −2
❿ −2, −8/3 ⓫ −5/2, −4/3 ⓬ −3, 8

Page 93

❶ 4, 4/5 ❷ −4/3, −7 ❸ −8, 9
❹ 9/5, 2/5 ❺ −3/5, 5/3 ❻ 1, 2/5
❼ 2/3, −9/4 ❽ 2, −7/5 ❾ 9/5, 4/3
❿ 7/2, 9/5 ⓫ 1, 7/3 ⓬ 7/2, 9

Page 94
❶ −1, 7/2 ❷ 3/2, 4/3 ❸ 2, 6
❹ −8, 4/5 ❺ −8/3, −1 ❻ −9, −3
❼ 3/2, 5/2 ❽ −5/3, −4 ❾ −7/4, 7/5
❿ −5/4, −1/2 ⓫ 2, −8 ⓬ 2, −3/5

Page 95
❶ 1, −1/2 ❷ −8, −9 ❸ 2, 5
❹ −1, 5/2 ❺ −3/2, 5/3 ❻ 8/5, 8/5
❼ −3/4, 7/3 ❽ −4, 9/4 ❾ −3/2, −1/2
❿ −2, −6 ⓫ −5/3, −3/2 ⓬ 5, −7/4

Page 96
❶ 7/4, 1/5 ❷ 8/3, −2 ❸ −9, 3
❹ 4/3, −4 ❺ 3, −7 ❻ −4/3, −3/2
❼ 5/4, −7/4 ❽ −8/3, −3 ❾ 9/5, −7/4
❿ −2/3, −3/2 ⓫ 1, 3/4 ⓬ −1, 7

Page 97
❶ 6/5, 3 ❷ 9/5, 2/3 ❸ 1/2, 1/4
❹ −3, 7/2 ❺ −3/5, 1/2 ❻ 8, 1/5
❼ −5/2, 1/3 ❽ 7, 2/5 ❾ −4/3, 9
❿ −3/2, 4/5 ⓫ −3/4, 4/5 ⓬ −2, 8

Page 98
❶ −1/2, −9/2 ❷ 4, 1 ❸ −2, −2
❹ −5/4, 3/4 ❺ 3/2, 1/4 ❻ −7/3, −7
❼ −9/4, −2/5 ❽ 7/4, 4/3 ❾ −1, −5/2
❿ −1/3, 7/2 ⓫ 3/2, −5/3 ⓬ −5, 3/4

Page 99
❶ 3/2, −9/5 ❷ −5, −5 ❸ −9/5, 2
❹ −8, −9/4 ❺ 8/3, −4 ❻ −1/5, −3
❼ −4/5, 7/2 ❽ −6, 8/5 ❾ −5/2, −1/4
❿ −4, −5/4 ⓫ 9/2, −1/4 ⓬ 2/5, −1/4

Page 100
❶ −2/5, 8 ❷ 1, 1 ❸ −3, 8/5
❹ 2, 5/3 ❺ 2, −9/4 ❻ −9, −8/5
❼ 1/4, 9 ❽ 7, −4/3 ❾ −9/2, −5
❿ −3/2, −9/4 ⓫ −6/5, 4/5 ⓬ 2/3, 7/3

Page 101
❶ 5/4, 3/5 ❷ 1/5, 1/3 ❸ −2/5, −7/3
❹ −5/3, −3 ❺ 1, −9/5 ❻ −6/5, 8
❼ 9/4, −1 ❽ 8/5, −3/4 ❾ 2, 1/3
❿ 3/4, −9/4 ⓫ 3, 2 ⓬ 2/5, 6/5

Page 102
❶ 3/4, −1/4 ❷ 3/5, −3/2 ❸ 1/5, 9/2
❹ −1/2, 1/3 ❺ −7, 4/3 ❻ 4, 8
❼ −7/3, −1 ❽ 3/2, 3/4 ❾ 2/5, −1/5
❿ 1/2, 7/5 ⓫ −7/4, 3/2 ⓬ 3/2, −6/5

Page 103
❶ −2, −1/2 ❷ −3/2, −4/3 ❸ −4/3, −2
❹ 1, 7/5 ❺ −1, 3/5 ❻ 9/2, 2
❼ 7/3, −5/2 ❽ −8, −3/5 ❾ 1/5, −6
❿ 4, 3/5 ⓫ 5/2, −1 ⓬ −7/4, 2

Page 104
❶ 1, 4 ❷ −1/5, 6/5 ❸ −3/2, −9/5
❹ 2, 2/5 ❺ −9/4, 4 ❻ −5, −2/5
❼ 8/3, 9/4 ❽ −1, 7/3 ❾ −7/4, −1/2
❿ −1/4, −6 ⓫ 5/4, −3/5 ⓬ 2/3, −5

Page 105
❶ 2, 5/4 ❷ −9/2, 9/5 ❸ −1/2, −7/4
❹ 1, 1/5 ❺ −1/5, −1/2 ❻ 2, −2
❼ −8/3, −2 ❽ −2/3, 2 ❾ −1/2, −7/4
❿ 1, 2 ⓫ 3, −5 ⓬ 9/5, −2/3

Page 106
❶ 1/4, −4 ❷ −5/3, 9/2 ❸ 4, −1/2
❹ −6/5, −2 ❺ −4/3, 4 ❻ −3, 9/4
❼ −6, −1 ❽ 7/3, −3/2 ❾ 1/5, 8
❿ 9/4, 1 ⓫ −6, 3/2 ⓬ −1, 9/5

Page 107
❶ 2, 2/5 ❷ 6, 6/5 ❸ −1, −5/3
❹ 3/5, 2/3 ❺ 4/5, 1 ❻ 6/5, 3
❼ 3/4, 8/5 ❽ 3, 4/5 ❾ 2/3, 5/4
❿ 1/5, 3 ⓫ −1/5, 1/5 ⓬ −3, −7/3

Page 108
❶ −9/4, 5 ❷ 4, 2 ❸ 8, −1
❹ −4, 4 ❺ 1, 3/4 ❻ 1/4, −1
❼ −3/5, 3 ❽ −7/3, 8/5 ❾ −1, 3/5
❿ 4/5, −8 ⓫ −3/5, −2/5 ⓬ −3, 2/5

Chapter 5 Answers:

Page 112

❶ $-7/5 \pm \sqrt{39}/5$ ❷ $-2 \pm \sqrt{5}$ ❸ $-5/6 \pm \sqrt{85}/6$

❹ $3 \pm \sqrt{10}$ ❺ $1/5 \pm \sqrt{41}/5$ ❻ $-8, -1$

❼ $-3/5, 2$ ❽ $-1, 1/4$ ❾ $-9/5 \pm 2\sqrt{14}/5$

❿ $5/4 \pm \sqrt{29}/4$ ⓫ $15/2 \pm 3\sqrt{21}/2$ ⓬ $-7/3 \pm \sqrt{55}/3$

⓭ $-5/2 \pm \sqrt{5}$ ⓮ $-11/8 \pm \sqrt{57}/8$ ⓯ $9/5 \pm \sqrt{46}/5$

⓰ $7 \pm 2\sqrt{14}$ ⓱ $-3/2, 1/2$ ⓲ $-1/2 \pm \sqrt{19}/2$

Page 113

❶ $4 \pm \sqrt{19}$ ❷ $-1 \pm 3\sqrt{2}/2$ ❸ $-6 \pm \sqrt{34}$

❹ $5/2 \pm \sqrt{26}/2$ ❺ $11/8 \pm 3\sqrt{17}/8$ ❻ $-3, 2/5$

❼ $-1 \pm \sqrt{22}/2$ ❽ $7/6 \pm \sqrt{97}/6$ ❾ $6 \pm 2\sqrt{11}$

❿ $4 \pm \sqrt{74}/2$ ⓫ $9 \pm 2\sqrt{21}$ ⓬ $-13/8 \pm 3\sqrt{33}/8$

⓭ $-2 \pm \sqrt{70}/5$ ⓮ $-9, -1/2$ ⓯ $8 \pm \sqrt{71}$

⓰ $-6/5 \pm \sqrt{41}/5$ ⓱ $-7/2 \pm 3\sqrt{7}/2$ ⓲ $-7/2, 1/2$

Page 114

❶ $2 \pm \sqrt{2}/2$ ❷ $-4 \pm 5\sqrt{2}/2$ ❸ $3 \pm \sqrt{26}/2$

❹ $1/2 \pm \sqrt{17}/2$ ❺ $-3, -1$ ❻ $3 \pm \sqrt{10}$

❼ $-2 \pm \sqrt{7}$ ❽ $-1/2 \pm \sqrt{21}/2$ ❾ $3/2 \pm 9\sqrt{5}/10$

❿ $5 \pm \sqrt{31}$ ⓫ $3 \pm 2\sqrt{3}$ ⓬ $-2, 1/5$

⓭ $-9/4 \pm 3\sqrt{17}/4$ ⓮ $7/5 \pm \sqrt{79}/5$ ⓯ $-9 \pm \sqrt{89}$

⓰ $7/4 \pm \sqrt{77}/4$ ⓱ $-3 \pm \sqrt{30}/2$ ⓲ $8 \pm \sqrt{69}$

Page 115

❶ $-3/2 \pm \sqrt{11}/2$ ❷ $-4 \pm \sqrt{15}$ ❸ $5/4 \pm \sqrt{17}/4$

❹ $-4/5 \pm \sqrt{61}/5$ ❺ $-1 \pm \sqrt{7}/2$ ❻ $5/3 \pm \sqrt{46}/3$

❼ $-9/5 \pm 4\sqrt{6}/5$ ❽ $2 \pm \sqrt{23}/2$ ❾ $7/4 \pm \sqrt{13}/4$

❿ $-2 \pm 2\sqrt{15}/3$ ⓫ $3 \pm 5\sqrt{3}/3$ ⓬ $-1/2 \pm \sqrt{5}/2$

⓭ $4 \pm \sqrt{66}/2$ ⓮ $-13/8 \pm \sqrt{89}/8$ ⓯ $17/6 \pm 5\sqrt{13}/6$

⓰ $1, 5/3$ ⓱ $7/4 \pm \sqrt{21}/4$ ⓲ $1/2, 9/2$

Page 116

❶ $-7/3 \pm \sqrt{22}/3$ ❷ $5/3 \pm \sqrt{22}/3$ ❸ $-3/4 \pm \sqrt{37}/4$

❹ $5/2 \pm \sqrt{57}/2$ ❺ $-4/5 \pm \sqrt{26}/5$ ❻ $1/3 \pm 2\sqrt{7}/3$

❼ $9/5 \pm \sqrt{41}/5$ ❽ $-1 \pm \sqrt{2}$ ❾ $-1/4, 5$

❿ $-1/2 \pm \sqrt{10}/2$ ⓫ $-5, -1$ ⓬ $-8 \pm 6\sqrt{2}$

⓭ $-1 \pm \sqrt{5}$ ⓮ $-7/2 \pm \sqrt{37}/2$ ⓯ $7/4 \pm \sqrt{77}/4$

⓰ $-4, 1/2$ ⓱ $-2 \pm \sqrt{14}/2$ ⓲ $-9/2 \pm \sqrt{85}/2$

Page 117

❶ $5/4 \pm \sqrt{21}/4$ ❷ $7/4 \pm \sqrt{13}/4$ ❸ $2 \pm \sqrt{7}$

❹ $3/2 \pm \sqrt{17}/2$ ❺ $-9/4 \pm 3\sqrt{13}/4$ ❻ $8 \pm 6\sqrt{2}$

❼ $3 \pm 4\sqrt{6}/3$ ❽ $1, 2$ ❾ $9/8 \pm \sqrt{17}/8$

❿ $-1/3, 4$ ⓫ $-1/4 \pm \sqrt{17}/4$ ⓬ $8 \pm \sqrt{65}$

⓭ $-3/2 \pm \sqrt{69}/6$ ⓮ $8/5 \pm 2\sqrt{6}/5$ ⓯ $8/5 \pm \sqrt{89}/5$

⓰ $2 \pm \sqrt{14}/2$ ⓱ $-3/2 \pm 9\sqrt{5}/10$ ⓲ $-7/4 \pm \sqrt{57}/4$

Page 118

❶ $-1/10 \pm \sqrt{41}/10$ ❷ $11/10 \pm 3\sqrt{29}/10$ ❸ $11/6 \pm \sqrt{61}/6$

❹ $-5 \pm \sqrt{21}$ ❺ $-1/4, 3$ ❻ $-2/5 \pm 2\sqrt{6}/5$

❼ $5 \pm \sqrt{86}/2$ ❽ $9/2 \pm \sqrt{67}/2$ ❾ $-7 \pm 5\sqrt{2}$

❿ $-3/2 \pm 3\sqrt{5}/2$ ⓫ $9/5 \pm 3\sqrt{14}/5$ ⓬ $1 \pm \sqrt{55}/5$

⓭ $5/2 \pm \sqrt{29}/2$ ⓮ $-7/2 \pm \sqrt{31}/2$ ⓯ $-1/4 \pm \sqrt{57}/4$

⓰ $2 \pm 2\sqrt{3}$ ⓱ $-1 \pm \sqrt{10}/2$ ⓲ $9/4 \pm \sqrt{73}/4$

Page 119

❶ $3/2 \pm \sqrt{17}/2$ ❷ $-4, -1$ ❸ $-4 \pm \sqrt{82}/2$

❹ $-4 \pm \sqrt{46}/2$ ❺ $-8 \pm \sqrt{67}$ ❻ $-9/4 \pm \sqrt{57}/4$

❼ $-5/2 \pm \sqrt{57}/2$ ❽ $7/4 \pm 3\sqrt{5}/4$ ❾ $-9/2 \pm \sqrt{69}/2$

❿ $2/5 \pm \sqrt{19}/5$ ⓫ $-1, -1/4$ ⓬ $-2 \pm 4\sqrt{5}/5$

⓭ $5/2 \pm \sqrt{53}/2$ ⓮ $-5/2 \pm \sqrt{39}/2$ ⓯ $-5/2 \pm \sqrt{41}/2$

⓰ $-2, -1/2$ ⓱ $7/2 \pm \sqrt{55}/2$ ⓲ $-3 \pm \sqrt{11}$

Page 120

❶ $-2/3 \pm \sqrt{13}/3$ ❷ $-17/2 \pm 3\sqrt{33}/2$ ❸ $-9, -1$

❹ $-2 \pm 2\sqrt{15}/3$ ❺ $-4/3, -1$ ❻ $-9/2 \pm 3\sqrt{7}/2$

❼ $8/3 \pm \sqrt{61}/3$ ❽ $4/5 \pm \sqrt{41}/5$ ❾ $-5/3 \pm \sqrt{19}/3$

❿ $-11/8 \pm \sqrt{57}/8$ ⓫ $-1 \pm \sqrt{30}/3$ ⓬ $4 \pm 3\sqrt{6}/2$

⓭ $-5/4 \pm \sqrt{21}/4$ ⓮ $-2 \pm \sqrt{95}/5$ ⓯ $-3/4 \pm \sqrt{65}/4$

⓰ $8/5 \pm \sqrt{39}/5$ ⓱ $1/5, 3$ ⓲ $-6 \pm \sqrt{38}$

Page 121

❶ $-7/2 \pm \sqrt{53}/2$ ❷ $-7/2 \pm \sqrt{33}/2$ ❸ $-2, 1/5$

❹ $11/2 \pm \sqrt{93}/2$ ❺ $7/4 \pm \sqrt{21}/4$ ❻ $7/4 \pm \sqrt{77}/4$

❼ $9 \pm \sqrt{83}$ ❽ $-1 \pm \sqrt{30}/5$ ❾ $-3 \pm \sqrt{10}$

❿ $-8 \pm 6\sqrt{2}$ ⓫ $-3, 1/3$ ⓬ $9/4 \pm \sqrt{77}/4$

⓭ $-9 \pm 6\sqrt{2}$ ⓮ $-13/2 \pm 3\sqrt{21}/2$ ⓯ $3/4 \pm 3\sqrt{5}/4$

⓰ $-5 \pm \sqrt{19}$ ⓱ $13/6 \pm \sqrt{61}/6$ ⓲ $-8/5 \pm \sqrt{59}/5$

Page 122

❶ $9 \pm \sqrt{86}$ ❷ $-5, -1/3$ ❸ $9/2 \pm \sqrt{97}/2$

❹ $3/2 \pm \sqrt{41}/2$ ❺ $-2 \pm \sqrt{5}$ ❻ $8/5 \pm \sqrt{94}/5$

❼ $5/8 \pm \sqrt{57}/8$ ❽ $-2 \pm \sqrt{14}/2$ ❾ $1/4 \pm \sqrt{65}/4$

❿ $1 \pm \sqrt{10}/5$ ⓫ $-3 \pm \sqrt{87}/3$ ⓬ $5/6 \pm \sqrt{97}/6$

⓭ $-1 \pm \sqrt{2}$ ⓮ $-9/2 \pm 3\sqrt{5}/2$ ⓯ $-4 \pm 3\sqrt{6}/2$

⓰ $-7/5 \pm \sqrt{19}/5$ ⓱ $6 \pm \sqrt{31}$ ⓲ $-7/2 \pm \sqrt{57}/2$

Page 123

❶ $1, 3/2$ ❷ $10/3 \pm \sqrt{82}/3$ ❸ $-5/2 \pm \sqrt{5}/2$

❹ $4 \pm \sqrt{13}$ ❺ $-7/4 \pm \sqrt{77}/4$ ❻ $3/5, 2$

❼ $-3 \pm \sqrt{15}$ ❽ $1 \pm \sqrt{5}$ ❾ $19/10 \pm 3\sqrt{29}/10$

❿ $1/6 \pm \sqrt{61}/6$ ⓫ $-13/6 \pm \sqrt{97}/6$ ⓬ $3/5 \pm 3\sqrt{6}/5$

⓭ $6/5 \pm \sqrt{31}/5$ ⓮ $-1 \pm \sqrt{2}/2$ ⓯ $-15/4 \pm 3\sqrt{33}/4$

⓰ $-1/4, 2$ ⓱ $5/4 \pm \sqrt{53}/4$ ⓲ $1/2 \pm \sqrt{3}/2$

Page 124

❶ $5/4 \pm \sqrt{13}/4$ ❷ $-5/8 \pm \sqrt{57}/8$ ❸ $-2/3, 3$

❹ $3/2 \pm \sqrt{6}/2$ ❺ $-2/3 \pm 2\sqrt{7}/3$ ❻ $-5 \pm \sqrt{21}$

❼ $3/2 \pm 3\sqrt{5}/10$ ❽ $-7/2 \pm \sqrt{85}/2$ ❾ $-8 \pm \sqrt{66}$

❿ $19/2 \pm 5\sqrt{13}/2$ ⓫ $-10 \pm 7\sqrt{2}$ ⓬ $1 \pm 2\sqrt{10}/5$

⓭ $1 \pm \sqrt{2}/2$ ⓮ $1, 2$ ⓯ $-4/5, 1$

⓰ $-5/2 \pm \sqrt{37}/2$ ⓱ $2 \pm \sqrt{10}$ ⓲ $-9/4 \pm \sqrt{65}/4$

Page 125

❶ $-4/3 \pm 2\sqrt{7}/3$ ❷ $1/2, 7/2$ ❸ $-1/2 \pm \sqrt{17}/2$

❹ $-8 \pm 3\sqrt{7}$ ❺ $5/2 \pm \sqrt{21}/2$ ❻ $-11/6 \pm \sqrt{85}/6$

❼ $-3 \pm \sqrt{11}$ ❽ $-9, 1$ ❾ $1/4 \pm \sqrt{5}/4$

❿ $-2 \pm \sqrt{5}$ ⓫ $-1 \pm \sqrt{6}$ ⓬ $-1/6 \pm \sqrt{85}/6$

⓭ $1/2, 3$ ⓮ $9/4 \pm \sqrt{57}/4$ ⓯ $2 \pm \sqrt{10}/2$

⓰ $5/6 \pm \sqrt{85}/6$ ⓱ $-1/2, 9$ ⓲ $-6/5 \pm \sqrt{21}/5$

Page 126

❶ $5/3 \pm \sqrt{34}/3$ ❷ $7/5 \pm \sqrt{34}/5$ ❸ $1, 7/3$

❹ $-3 \pm 5\sqrt{3}/3$ ❺ $-1 \pm \sqrt{2}$ ❻ $-8/3 \pm \sqrt{73}/3$

❼ $2 \pm \sqrt{15}/2$ ❽ $-8/3, 1$ ❾ $-2, -4/5$

❿ $-10 \pm \sqrt{94}$ ⓫ $-5/2 \pm \sqrt{37}/2$ ⓬ $-17/4 \pm 3\sqrt{33}/4$

⓭ $-2 \pm \sqrt{5}$ ⓮ $3/2 \pm \sqrt{2}/2$ ⓯ $1/4 \pm \sqrt{41}/4$

⓰ $5/4 \pm \sqrt{33}/4$ ⓱ $7/4 \pm 3\sqrt{5}/4$ ⓲ $-3/2 \pm 3\sqrt{5}/10$

Page 127

❶ $-11/4 \pm \sqrt{97}/4$ ❷ $-1, 1/2$ ❸ $-1/2, 5$

❹ $2 \pm \sqrt{23}/2$ ❺ $-9 \pm \sqrt{82}$ ❻ $3 \pm \sqrt{78}/3$

❼ $1 \pm 3\sqrt{2}/2$ ❽ $11/4 \pm \sqrt{97}/4$ ❾ $-1/4 \pm \sqrt{33}/4$

❿ $-4, 1/2$ ⓫ $7 \pm \sqrt{51}$ ⓬ $7/4 \pm \sqrt{17}/4$

⓭ $-2, 1/5$ ⓮ $-8 \pm \sqrt{67}$ ⓯ $-2 \pm \sqrt{26}/2$

⓰ $-9/2 \pm \sqrt{65}/2$ ⓱ $10 \pm 6\sqrt{3}$ ⓲ $5/2 \pm \sqrt{41}/2$

Page 128

❶ $7/3 \pm 2\sqrt{10}/3$ ❷ $-3/2, -1$ ❸ $-6/5 \pm \sqrt{71}/5$

❹ $-3/5 \pm 3\sqrt{6}/5$ ❺ $19/2 \pm 3\sqrt{41}/2$ ❻ $-2/3, 3$

❼ $-7/8 \pm \sqrt{97}/8$ ❽ $13/4 \pm 3\sqrt{17}/4$ ❾ $11/10 \pm \sqrt{41}/10$

❿ $1/2 \pm \sqrt{69}/6$ ⓫ $-5/4 \pm \sqrt{17}/4$ ⓬ $-4, -1/4$

⓭ $17/10 \pm 3\sqrt{41}/10$ ⓮ $-1/6 \pm \sqrt{73}/6$ ⓯ $-4 \pm \sqrt{11}$

⓰ $-17/6 \pm 5\sqrt{13}/6$ ⓱ $13/8 \pm \sqrt{89}/8$ ⓲ $1, 7$

Page 129

❶ $-3 \pm \sqrt{78}/3$ ❷ $4 \pm 2\sqrt{3}$ ❸ $9/5 \pm \sqrt{66}/5$

❹ $7/4 \pm \sqrt{85}/4$ ❺ $1/4 \pm \sqrt{21}/4$ ❻ $2 \pm \sqrt{2}$

❼ $4 \pm \sqrt{19}$ ❽ $1 \pm \sqrt{5}$ ❾ $-2/3 \pm \sqrt{10}/3$

❿ $4 \pm \sqrt{11}$ ⓫ $-9/2 \pm \sqrt{93}/2$ ⓬ $3/10 \pm \sqrt{69}/10$

⓭ $-7/5 \pm 2\sqrt{21}/5$ ⓮ $-7/4 \pm \sqrt{29}/4$ ⓯ $9/2 \pm \sqrt{71}/2$

⓰ $5 \pm 4\sqrt{2}$ ⓱ $-4 \pm 3\sqrt{2}$ ⓲ $3/2 \pm 7\sqrt{5}/10$

Page 130

❶ $-1, 9/4$ ❷ $-4 \pm 2\sqrt{3}$ ❸ $-3, -2$

❹ $-4/5 \pm \sqrt{61}/5$ ❺ $-8/3 \pm 2\sqrt{19}/3$ ❻ $9/2 \pm \sqrt{97}/2$

❼ $-5/4 \pm \sqrt{41}/4$ ❽ $-7/2 \pm \sqrt{85}/2$ ❾ $-3/2 \pm \sqrt{21}/2$

❿ $-1, 4/3$ ⓫ $9/8 \pm \sqrt{33}/8$ ⓬ $7/3 \pm \sqrt{58}/3$

⓭ $9/5 \pm 3\sqrt{14}/5$ ⓮ $5/2 \pm \sqrt{33}/2$ ⓯ $11/2 \pm \sqrt{89}/2$

⓰ $-1, -1/4$ ⓱ $-3/10 \pm \sqrt{89}/10$ ⓲ $-9/2 \pm \sqrt{57}/2$

Page 131

❶ $15/4 \pm 3\sqrt{17}/4$ ❷ $-7/4 \pm \sqrt{65}/4$ ❸ $-5/2 \pm \sqrt{29}/2$

❹ $-8/5 \pm \sqrt{89}/5$ ❺ $-4/3 \pm \sqrt{37}/3$ ❻ $5 \pm 2\sqrt{6}$

❼ $-1/4 \pm \sqrt{21}/4$ ❽ $-4 \pm \sqrt{15}$ ❾ $-6 \pm \sqrt{39}$

❿ $1/8 \pm \sqrt{65}/8$ ⓫ $-1/2, 5/2$ ⓬ $-2 \pm \sqrt{6}$

⓭ $2/3, 1$ ⓮ $6 \pm \sqrt{34}$ ⓯ $10/3 \pm \sqrt{85}/3$

⓰ $17/10 \pm 3\sqrt{41}/10$ ⓱ $8/5 \pm \sqrt{19}/5$ ⓲ $-9/5 \pm 4\sqrt{6}/5$

Chapter 6 Answers:

Page 134

❶ $3 \pm 2\sqrt{13}$ ❷ 61/20 ❸ −2, −1/4

❹ 19/24 ❺ −1/24 ❻ −2/25

❼ $11/36 \pm \sqrt{73}/36$ ❽ 8/9 ❾ −25/32

❿ 2 ⓫ $-9/4 \pm \sqrt{15}/4$ ⓬ 8/5

⓭ 1/7 ⓮ 20 ⓯ $-7/36 \pm \sqrt{13}/36$

Page 135

❶ −13/5 ❷ −5/3, −4/3 ❸ $-9/4 \pm \sqrt{41}/4$

❹ −1/12 ❺ −2/5 ❻ −11/17

❼ 34 ❽ $-9/4 \pm 3\sqrt{17}/4$ ❾ $2/3 \pm \sqrt{5}/15$

❿ $-31/8 \pm 3\sqrt{57}/8$ ⓫ 0 ⓬ 0

⓭ $11/2 \pm \sqrt{57}/6$ ⓮ 67/61 ⓯ −55/36

Page 136

❶ $3/2 \pm \sqrt{3}/3$ ❷ 18/47 ❸ 0

❹ −1/2 ❺ $-1 \pm 2\sqrt{10}$ ❻ −48/41

❼ −9/20 ❽ 1/7, 3/7 ❾ −8/5

❿ $-1 \pm \sqrt{31}$ ⓫ 20/41 ⓬ $-7/8 \pm \sqrt{97}/8$

⓭ 21/23 ⓮ 4 ⓯ −13/11

Page 137

❶ 4/21, 1 ❷ −1, −1/2 ❸ $-1/3 \pm \sqrt{5}/15$

❹ $-3/10 \pm \sqrt{34}/10$ ❺ 3 ❻ −7/3, 1

❼ −1, −13/42 ❽ −1, 3 ❾ 10/23

❿ 38/29 ⓫ 53/25 ⓬ $-4/7 \pm \sqrt{5}/7$

⓭ −18/5 ⓮ 1, 43/40 ⓯ $-10/3 \pm \sqrt{91}/3$

Page 138

❶ 1/4 ❷ $-3/2 \pm \sqrt{22}/2$ ❸ −40/27

❹ 1, 7/4 ❺ $-2 \pm 2\sqrt{3}$ ❻ −7, −5

❼ −5/8 ❽ 5/13 ❾ $11/8 \pm \sqrt{65}/8$

❿ $-1/9 \pm \sqrt{13}/9$ ⓫ 4/3, 3 ⓬ $3/2 \pm \sqrt{11}/6$

⓭ 13/12 ⓮ −1/7 ⓯ 21/2

Page 139

❶ $2/3 \pm \sqrt{43}/6$ ❷ $-5/6 \pm \sqrt{17}/6$ ❸ 18/35

❹ $7/12 \pm \sqrt{13}/12$ ❺ $-7/3 \pm 2\sqrt{19}/3$ ❻ 29/14

❼ $-5/2 \pm \sqrt{89}/2$ ❽ −23/59 ❾ −11, 4

❿ 51/14 ⓫ −5/3 ⓬ −2/3

⓭ 29/52 ⓮ −44/37 ⓯ $1/10 \pm \sqrt{11}/10$

Page 140

❶ $3/2 \pm \sqrt{15}/2$ ❷ $-11/14 \pm \sqrt{17}/14$ ❸ −14/29

❹ $-17/4 \pm \sqrt{89}/4$ ❺ $-2/7 \pm \sqrt{74}/7$ ❻ $-5/4 \pm \sqrt{57}/12$

❼ 7/19 ❽ $1/2 \pm \sqrt{5}/2$ ❾ 25/17

❿ $3/4 \pm 3\sqrt{41}/4$ ⓫ 1/3, 4/9 ⓬ $21/8 \pm \sqrt{73}/8$

⓭ −3/2 ⓮ $17/18 \pm \sqrt{97}/18$ ⓯ $1/8 \pm \sqrt{73}/8$

Page 141

❶ $1/4 \pm \sqrt{33}/4$ ❷ $21/8 \pm \sqrt{41}/8$ ❸ $34/63 \pm 2\sqrt{37}/63$

❹ $1/3 \pm 2\sqrt{6}/9$ ❺ 9/25 ❻ 3/2

❼ 5/59 ❽ −59/66 ❾ 5/13

❿ $3 \pm \sqrt{3}$ ⓫ −1/2, −1/8 ⓬ −7/9

⓭ 17/4 ⓮ $-1 \pm 2\sqrt{6}/3$ ⓯ −27/23

Page 142

❶ $2 \pm 2\sqrt{10}$ ❷ 1/2 ❸ $-5/16 \pm \sqrt{57}/16$

❹ $-4/3 \pm \sqrt{17}/3$ ❺ 1/2 ❻ −27/23

❼ −3/2, 7 ❽ −1/47 ❾ 4/39

❿ $40/21 \pm 2\sqrt{43}/21$ ⓫ $1/3 \pm \sqrt{19}/3$ ⓬ −5/4, −2/3

⓭ $-11/12 \pm \sqrt{61}/12$ ⓮ 5/3, 4 ⓯ 12/35

Page 143

❶ $-8/3 \pm \sqrt{31}/3$ ❷ 50/9 ❸ $11/12 \pm \sqrt{85}/12$

❹ 1/10 ❺ 1/7 ❻ 28/15

❼ 17/47 ❽ $5/4 \pm \sqrt{6}/4$ ❾ 29/36

❿ $1/4 \pm \sqrt{33}/4$ ⓫ −1/32 ⓬ −5/6

⓭ $-13/3 \pm \sqrt{43}/3$ ⓮ $1/2 \pm \sqrt{65}/2$ ⓯ −11/21

Page 144

❶ 65/69 ❷ 8/17 ❸ $5/6 \pm \sqrt{3}/2$

❹ 6/61 ❺ $\pm \sqrt{19}$ ❻ 1/2

❼ $1/10 \pm \sqrt{41}/10$ ❽ 7/19 ❾ $1/3 \pm \sqrt{2}/3$

❿ 6/7 ⓫ $2 \pm \sqrt{3}$ ⓬ −16

⓭ $-1/20 \pm \sqrt{41}/20$ ⓮ −9/2, −1/3 ⓯ 6/13

Page 145

❶ 6 ❷ −5/2 ❸ $-13/5 \pm 2\sqrt{41}/5$

❹ 20/27 ❺ $-9/10 \pm \sqrt{17}/10$ ❻ −56/11

❼ 1/2, 8/3 ❽ 9/11 ❾ $\pm \sqrt{6}/2$

❿ −17/15 ⓫ $-6/7 \pm 2\sqrt{2}/7$ ⓬ 17/44

⓭ 11/23 ⓮ $-13/4 \pm \sqrt{37}/4$ ⓯ 63/4

Page 146

❶ $-8/3$ ❷ $21/19$ ❸ $-7/9$

❹ $1 \pm \sqrt{39}/6$ ❺ $-36/55$ ❻ -1

❼ $-3 \pm 2\sqrt{7}$ ❽ $3/13$ ❾ $-9/2 \pm 3\sqrt{17}/2$

❿ $\pm 2\sqrt{3}/3$ ⓫ $1/3 \pm \sqrt{46}/6$ ⓬ $35/73$

⓭ $-7/2, 1$ ⓮ $33/25$ ⓯ $-11/4 \pm \sqrt{65}/4$

Page 147

❶ $1/5 \pm \sqrt{6}/5$ ❷ $-5/2$ ❸ $-21/22$

❹ $1/4 \pm \sqrt{21}/4$ ❺ $1 \pm \sqrt{11}$ ❻ $33/20 \pm \sqrt{89}/20$

❼ $1/6 \pm \sqrt{7}/3$ ❽ $53/39$ ❾ $7/12 \pm \sqrt{97}/12$

❿ $4/3, 5/2$ ⓫ $-3/14 \pm \sqrt{53}/42$ ⓬ $\pm \sqrt{13}$

⓭ $27/19$ ⓮ $-2/3 \pm \sqrt{22}/3$ ⓯ $1/24, 1/2$

Page 148

❶ $-1/3, 2/3$ ❷ $1 \pm \sqrt{58}/4$ ❸ $-34/15, -1$

❹ $5/2 \pm \sqrt{14}$ ❺ $\pm \sqrt{6}/3$ ❻ $-4/3$

❼ $6/5$ ❽ $-1/12 \pm \sqrt{73}/12$ ❾ $1/2 \pm \sqrt{97}/2$

❿ $1/15$ ⓫ $-6/7, 1/3$ ⓬ $-83/60$

⓭ $-17/21 \pm \sqrt{79}/21$ ⓮ $3/2 \pm \sqrt{11}/2$ ⓯ $17/15$

Page 149

❶ $-3/8, -1/4$ ❷ $-5/8 \pm \sqrt{17}/8$ ❸ $\pm \sqrt{6}$

❹ $-13/17$ ❺ $-30/23$ ❻ $-3/2 \pm \sqrt{41}/6$

❼ $-3, 1/2$ ❽ $\pm \sqrt{11}$ ❾ $7/13$

❿ $-8/9$ ⓫ $3/4 \pm \sqrt{69}/4$ ⓬ $40/17$

⓭ $-3/2 \pm \sqrt{29}/2$ ⓮ $5/14, 1/2$ ⓯ $-4/3$

Page 150

❶ $1/3$ ❷ $-17/12$ ❸ $5/4 \pm \sqrt{33}/4$

❹ $-2, 7$ ❺ $-34/35$ ❻ $-29/18$

❼ $-5/7 \pm \sqrt{58}/14$ ❽ $3/2 \pm \sqrt{57}/6$ ❾ $-1/12 \pm \sqrt{61}/12$

❿ $-62/9$ ⓫ $-17/12 \pm \sqrt{97}/12$ ⓬ $7/6 \pm \sqrt{13}/6$

⓭ $-26/27$ ⓮ $-5/14$ ⓯ $1/2, 5/8$

Page 151

❶ $-2 \pm \sqrt{21}/3$ ❷ $-79/31$ ❸ $2/7$

❹ $-5/2 \pm \sqrt{11}/2$ ❺ $-1/3 \pm \sqrt{19}/3$ ❻ $-1, 0$

❼ $2/3$ ❽ $\pm \sqrt{2}/2$ ❾ $9/11$

❿ $-5/4$ ⓫ $11/2 \pm \sqrt{89}/2$ ⓬ $1/2$

⓭ -1 ⓮ $1/10$ ⓯ $10/11$

Page 152

❶ $5/2$ ❷ $13/49$ ❸ $-32/33$

❹ $1/2 \pm \sqrt{6}/6$ ❺ $-83/15$ ❻ $-7/2, -1/2$

❼ $17/5 \pm 2\sqrt{61}/5$ ❽ $-14/3 \pm \sqrt{31}/3$ ❾ $-2/3 \pm \sqrt{22}/6$

❿ $7/2 \pm \sqrt{33}/6$ ⓫ $3/8 \pm \sqrt{65}/8$ ⓬ $23/24$

⓭ $-3/2 \pm \sqrt{14}/4$ ⓮ $7/6$ ⓯ $-4 \pm \sqrt{33}/2$

Page 153

❶ $-5/17$ ❷ $-15/8 \pm \sqrt{37}/8$ ❸ $5 \pm \sqrt{2}$

❹ $-9/8 \pm \sqrt{97}/8$ ❺ $3/2 \pm \sqrt{46}/4$ ❻ $9/14$

❼ $2/3$ ❽ $1/30 \pm \sqrt{61}/30$ ❾ $-11/16 \pm 3\sqrt{17}/16$

❿ $7/3$ ⓫ $1 \pm \sqrt{30}/3$ ⓬ $5/6 \pm \sqrt{5}/2$

⓭ $1, 3$ ⓮ $1/2, 19/24$ ⓯ $-11/2, -2$

Page 154

❶ $4/3 \pm 2\sqrt{10}/3$ ❷ $-6/13$ ❸ $3/11$

❹ $-1/6 \pm \sqrt{3}/2$ ❺ $1/8 \pm \sqrt{53}/8$ ❻ $-11/16 \pm \sqrt{33}/16$

❼ $-2 \pm \sqrt{2}/2$ ❽ $-15/14$ ❾ $4/3 \pm \sqrt{37}/6$

❿ $-7/20 \pm \sqrt{89}/20$ ⓫ $-6/17$ ⓬ $-9/2 \pm \sqrt{14}/2$

⓭ $-2/3 \pm \sqrt{46}/6$ ⓮ $\pm \sqrt{10}$ ⓯ $7/9$

Page 155

❶ -3 ❷ $-79/32$ ❸ $-3/4 \pm \sqrt{41}/4$

❹ $-11/17$ ❺ $28/31$ ❻ $-7/5, -2/5$

❼ $9/20$ ❽ $\pm \sqrt{6}/12$ ❾ $-9/43$

❿ $5 \pm \sqrt{19}$ ⓫ $3/10 \pm \sqrt{21}/70$ ⓬ $-44/41$

⓭ $7/2, 6$ ⓮ $-108/17$ ⓯ $7/6 \pm \sqrt{61}/6$

Page 156

❶ $-2/3 \pm \sqrt{13}/6$ ❷ $35/32$ ❸ $1/3, 2/3$

❹ $-1/15$ ❺ $-1/11$ ❻ $67/11$

❼ $-7/4, 0$ ❽ $1/3 \pm \sqrt{58}/6$ ❾ 22

❿ $12/31$ ⓫ $-1/2, 6$ ⓬ $-4 \pm \sqrt{22}$

⓭ $-26/3$ ⓮ $135/7$ ⓯ $9/2 \pm 3\sqrt{11}/2$

Page 157

❶ -5 ❷ $43/45$ ❸ 5

❹ ± 12 ❺ $3/2 \pm \sqrt{61}/2$ ❻ $-2/3$

❼ -2 ❽ $9/10 \pm \sqrt{11}/10$ ❾ $-2 \pm \sqrt{33}$

❿ $5/4 \pm \sqrt{15}/4$ ⓫ $-21/26$ ⓬ $-5/3, -2/3$

⓭ $25/4 \pm \sqrt{97}/4$ ⓮ $1/4$ ⓯ $5/8 \pm \sqrt{33}/24$

Chapter 7 Answers:

Page 163
❶ 4/3, 29/18 ❷ –12/83, –34/83 ❸ –21, 36
❹ –10/37, 26/37 ❺ –1/7, 16/21 ❻ –7/9, –8/9
❼ –5/13, 1/26 ❽ –5/33, –59/33 ❾ –7/3, 5/3
❿ –1/21, –19/21 ⓫ 23/61, –20/61 ⓬ 27/40, –31/20

Page 164
❶–3/7, –1/2 ❷ –15/2, 9/2 ❸ –4/3, 2
❹ 10/9, –13/9 ❺ –17/18, 23/36 ❻ –2/31, –15/31
❼ –70/27, –49/27 ❽ –26/47, –4/47 ❾ –24/7, –43/21
❿ 10/77, –8/11 ⓫ 8/5, 11/5 ⓬ –1, –7/3

Page 165
❶ 19/3, 13/3 ❷ 19/14, –7/2 ❸ –3/5, –2
❹ –11/6, –19/18 ❺ 9/14, –11/28 ❻ –11/79, –59/79
❼ –61/37, –33/37 ❽ 1/2, 1/2 ❾ 32/3, 25/3
❿ 2/5, –7/5 ⓫ 0, 3/7 ⓬ 1/12, –13/6

Page 166
❶ 1/21, 17/21 ❷ 52/79, 6/79 ❸ –49/81, –121/81
❹ –26/19, 25/19 ❺ –86/55, 21/55 ❻ –9/8, –1/8
❼ 47/76, –43/76 ❽ –1/4, 11/32 ❾ 7/8, 1/4
❿ –15/2, –7 ⓫ –1, –3 ⓬ –91/45, –14/5

Page 167
❶ 5/6, –1/9 ❷ 29/24, 5/6 ❸ –1/22, 59/44
❹ 44/65, –6/65 ❺ –7/41, –10/41 ❻ –8/13, –5/13
❼ –3/2, –11/6 ❽ –13/33, –2/11 ❾ 17, –10
❿ 3/16, –33/32 ⓫ 59/71, 66/71 ⓬ –63/52, –17/26

Page 168
❶ 69/62, 15/62 ❷ 0, –5/7 ❸ –37/21, –20/7
❹ –33/8, –13/4 ❺ 17/23, 39/23 ❻ –7/5, 11/10
❼ –26/27, 14/9 ❽ –43/2, –10 ❾ –1/4, –15/28
❿ –1/10, 37/40 ⓫ –3/35, 52/35 ⓬ –21/13, –6/13

Page 169
❶ 11/4, 13/4 ❷ –61/62, –69/62 ❸ –47/32, –35/32
❹ –5/24, –17/8 ❺ –5/9, –40/9 ❻ 16/51, 11/51
❼ –6, –27/4 ❽ 29/11, 15/11 ❾ 3/4, –3/4
❿ –23/50, –3/50 ⓫ 23/36, 5/8 ⓬ 43/50, 9/50

Page 170
❶ 109/12, 13/2 ❷ –26/19, –21/19 ❸ –5, 3/2
❹ –11/5, 26/15 ❺ –13/7, –9/14 ❻ 47/44, –5/44
❼ 7, 6 ❽ –10, –1 ❾ 8/3, 3
❿ 15/11, 73/33 ⓫ –17/19, 24/19 ⓬ –7/20, –37/20

Page 171
❶ 35/22, 16/11 ❷ 31/10, –17/5 ❸ 14, 31/2
❹ –23/91, 94/91 ❺ –2/7, 8/7 ❻ –29/13, 14/13
❼ –9/41, 24/41 ❽ 3, –13/3 ❾ 9/7, –1/7
❿ –1/2, –7/2 ⓫ –1/5, –29/35 ⓬ 60/29, 21/29

Page 172
❶ 16/11, 10/11 ❷ –46/41, –57/41 ❸ 11, 13/2
❹ –9/5, –2/5 ❺ 3/19, –56/19 ❻ 23/29, –36/29
❼ –4/43, –21/43 ❽ –75/8, –13/4 ❾ 3/2, –1
❿ 9/2, –3/2 ⓫ –17/8, –21/16 ⓬ 35/22, –1/22

Page 173
❶ 13/7, 69/28 ❷ –13/22, –59/44 ❸ 68, –30
❹ –1, –7/2 ❺ 61/36, –11/4 ❻ 13/34, 33/34
❼ –5/28, –39/28 ❽ –7/12, –5/12 ❾ –22/43, –71/43
❿ 97/89, 7/89 ⓫ –7/5, 1 ⓬ 11/21, 4/7

Page 174
❶ 7, –7 ❷ 5/92, –55/92 ❸ 17/7, –15/7
❹ 90/43, 83/43 ❺ 3/14, –19/14 ❻ –2/3, –2/3
❼ 33/32, 61/32 ❽ –2/11, 32/33 ❾ 128/19, 99/19
❿ 15/7, –19/14 ⓫ 1/2, –5/4 ⓬ –7/29, –17/87

Page 175
❶ 120, –16 ❷ –6, –19 ❸ –1/36, –11/12
❹ 55/12, –13/4 ❺ –5/6, –13/6 ❻ –23/13, –15/13
❼ –77/3, 49/3 ❽ –1/3, –10/3 ❾ –17/11, –54/11
❿ –59/25, 6/25 ⓫ 3, 2 ⓬ 4, –26/5

Page 176
❶ 7/2, –1/6 ❷ –3/5, –49/5 ❸ 11/8, –17/16
❹ 35/46, 91/46 ❺ –9/8, 5/8 ❻ –9/5, 12/5
❼ 3/19, –14/19 ❽ –59/48, –13/16 ❾ 26/55, 43/55
❿ 51/7, 1/7 ⓫ –3/7, 9/7 ⓬ –7/2, –9/2

Page 177
❶ 21/11, 38/11 ❷ 7/17, 50/51 ❸ 82/67, –48/67
❹ –49/43, 1/43 ❺ 10/37, 19/37 ❻ –1, 1
❼ 25/37, 72/37 ❽ –5/3, –3 ❾ 13/9, –14/9
❿ –13/10, –11/10 ⓫ –8/9, –7/9 ⓬ 27/31, –13/62

Page 178
❶ –46/49, –32/49 ❷ 13/10, –27/10 ❸ –5/14, –10/7
❹ 9/5, 13/5 ❺ 5/11, –29/44 ❻ 3/4, –5/4
❼ 6/37, –36/37 ❽ 6/5, –1/45 ❾ –77/38, 17/38
❿ –31/108,–43/108 ⓫ –59/94,–21/47 ⓬ –29/36, 4/9

Page 179
❶ 19/47, 6/47 ❷ –4/3, –1 ❸ 1/109, –25/109
❹ –29/6, 101/12 ❺ –13/23, –21/46 ❻ –45/2, –12
❼ 3/14, 20/21 ❽ –5/3, –4/3 ❾ 12/11, 38/33
❿ –23/15, –22/15 ⓫ 2/17, –1/153 ⓬ 2/13, 17/39

Page 180
❶ 27/8, –7/8 ❷ 1/4, –5/4 ❸ 9/31, –46/31
❹ –20/7, 23/7 ❺ –13/70, 53/70 ❻ –31/45, –1/5
❼ 8/5, –11/45 ❽ 1/3, 11/6 ❾ –1, –1
❿ –5/2, –3/2 ⓫ 2/11, 9/11 ⓬ –13/21, –10/7

Page 181
❶ –1/2, 1/4 ❷ 112/25, 119/25 ❸ –1/2, 0
❹ 1/15, –4/5 ❺ –16/39, –67/78 ❻ 13/12, –7/3
❼ 31/17, 33/17 ❽ –49/26, –6/13 ❾ 11/12, 7/4
❿ –17/3, –1/3 ⓫ –1/2, 5/2 ⓬ –16/15, 7/15

Page 182
❶ 9/13, 7/13 ❷ –27/43, –15/86 ❸ –2, 7/2
❹ 16/27, –10/9 ❺ –9, –7 ❻ 5/3, –1/3
❼ 86/45, 14/5 ❽ –3/62, –10/31 ❾ 1/2, 1/2
❿ –1/11, –9/11 ⓫ –21/71, 52/71 ⓬ –1/9, 11/3

Page 183
❶ 17/18, 23/18 ❷ 25/26, –11/26 ❸ –29/34, 4/17
❹ –37/9, –14/3 ❺ 29/64, –21/32 ❻ –3/4, –19/4
❼ 19/17, 7/17 ❽ –67/13, –17/13 ❾ 18/17, 25/17
❿ 36/53, –10/53 ⓫ –8/11, 13/11 ⓬ 7/9, –8/9

Page 184
❶ –20/13, 5/13 ❷ 48/11, –28/11 ❸ –1, –3
❹ 23/8, 3 ❺ –2/5, –1 ❻ –15/2, 37/6
❼ –8, 9/2 ❽ –96/23, 57/23 ❾ 5/3, 11/6
❿ 15/8, 71/8 ⓫ 4/3, –2/21 ⓬ 45/38, –39/76

Page 185
❶ 4/7, 3/2 ❷ –11/7, 19/7 ❸ 6/7, –55/21
❹ 55/7, 18/7 ❺ 21/20, 27/20 ❻ –81/22, 83/22
❼ 19/48, –55/96 ❽ 5/64, 11/8 ❾ 13/3, 15/2
❿ –5/3, 29/27 ⓫ 18/11, –13/11 ⓬ –31/35, 2/7

Page 186
❶ –5, –6 ❷ 11/32, 1/4 ❸ –9/29, 37/29
❹ 1/3, –5/9 ❺ –23/5, –1/5 ❻ 13/7, 5/7
❼ 22/15, –49/15 ❽ –39/5, –41/5 ❾ –2/11, –10/33
❿ 7/13, 4/13 ⓫ 33, –24 ⓬ –7/3, –2/3

Page 187
❶ –71/66, –3/22 ❷ –23/30, 13/30 ❸ 47/6, 19/6
❹ 17/6, –95/24 ❺ 2, 11/4 ❻ –1, –1/3
❼ 22/9, 29/9 ❽ 27, –32 ❾ 29/64, 5/8
❿ –47/27, 16/27 ⓫ –13/19,–31/19 ⓬ –48/17,–57/17

Page 188
❶ 37/88, –1/88 ❷ –40/49, 33/49 ❸ –17/26, 7/26
❹ –31/16, 3/32 ❺ –58/51, –89/51 ❻ –47/3, 26/3
❼ 8/37, –39/37 ❽ 10/43, –16/43 ❾ –30/23, –25/23
❿ –5/2, –9/2 ⓫ 1/2, –3/2 ⓬ 36/13, 29/13

Page 189
❶ –31/29, 39/29 ❷ –13/10, –2/5 ❸ –23/12, 65/24
❹ –9/14, –67/14 ❺ 4, –3 ❻ 3, –2/3
❼ –4/9, –52/63 ❽ –13/67, 43/67 ❾ –11/5, –3/5
❿ 7/33, 21/22 ⓫ 33/28, 1/4 ⓬ 2/5, –11/5

Page 190
❶ –2, 1 ❷ 9/29, 30/29 ❸ 3/22, –39/22
❹ 31/8, 45/8 ❺ 31/49, 1/14 ❻ –17/21, 10/7
❼ –28/43, 4/43 ❽ 0, 3 ❾ 21/13, 45/13
❿ –13, –5 ⓫ –1, 7 ⓬ 9/17, –5/17

Page 191
❶ –2, 0 ❷ 11/2, –3/2 ❸ 17/9, –29/18
❹ 13/9, –8/9 ❺ –1/10, 17/10 ❻ –59/16, –95/16
❼ 9/4, –1/4 ❽ 14/9, 25/9 ❾ 37/16, –65/16
❿ 1/5, –17/10 ⓫ 1/3, 0 ⓬ 7/38, –67/38

Page 192
❶ 1, 0 ❷ 20/67, 46/67 ❸ 1/10, –1/2
❹ –2, –1 ❺ 6/5, –53/45 ❻ –5, –44/3
❼ –5/7, –12/7 ❽ 1/4, 19/20 ❾ 25/16, 5/16
❿ 30/13,–36/13 ⓫ 35/24, 41/24 ⓬ –16/11, 2/11

Improve Your Math Fluency

www.improveyourmathfluency.com

Chris McMullen, Ph.D.

www.chrismcmullen.wordpress.com

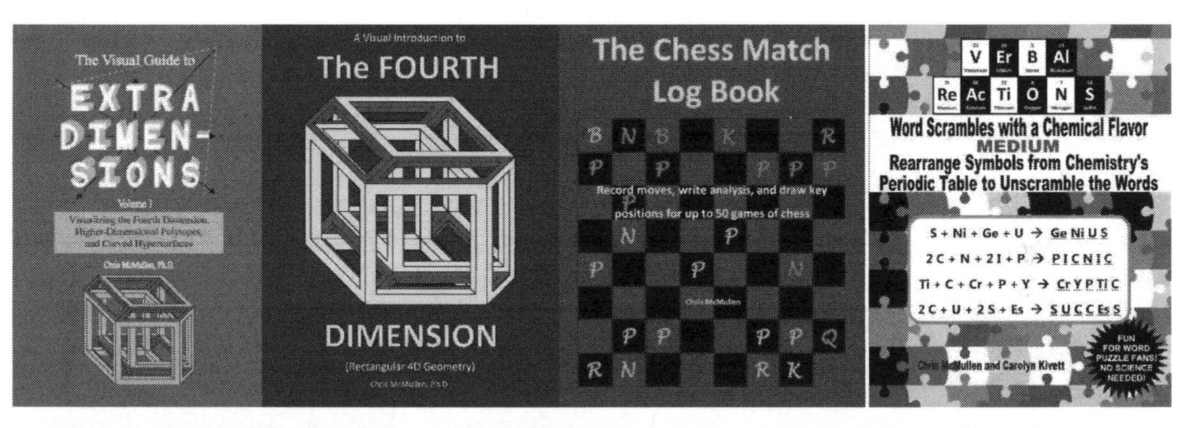

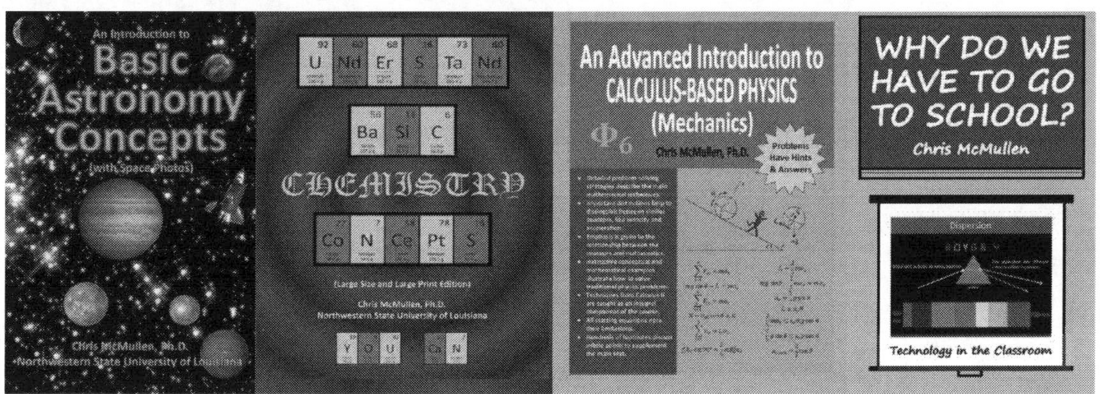

Made in the USA
Middletown, DE
24 June 2017